PHAENOMENA

Johns Hopkins
New Translations
from Antiquity

Aratus

PHAENOMENA

Translated, with an introduction and notes,
by Aaron Poochigian

The Johns Hopkins University Press
Baltimore

Printed in the United States of America on acid-free paper
9 8 7 6 5 4 3 2 1

The Johns Hopkins University Press
2715 North Charles Street
Baltimore, Maryland 21218-4363
www.press.jhu.edu

Library of Congress Cataloging-in-Publication Data

Aratus, Solensis.
[Phaenomena. English]
Phaenomena / Aratus ; translated, with an introduction and notes, by Aaron Poochigian.
p. cm.
"Includes a verse adaptation of Eudoxus' book of the same name"—Data view.
Includes bibliographical references.
ISBN-13: 978-0-8018-9465-7 (hardcover : alk. paper)
ISBN-10: 0-8018-9465-4 (hardcover : alk. paper)
ISBN-13: 978-0-8018-9466-4 (pbk. : alk. paper)
ISBN-10: 0-8018-9466-2 (pbk. : alk. paper)
1. Didactic poetry, Greek—Translations into English. 2. Astronomy, Ancient—Poetry. 3. Constellations—Poetry. 4. Planets—Poetry. 5. Weather—Poetry. I. Poochigian, Aaron, 1973– II. Eudoxus, of Cnidus, ca. 400–ca. 350 B.C. Phaenomena. III. Title.
QB802.A713 2010
520—dc22 2009038861

A catalog record for this book is available from the British Library.

All illustrations are from Hyginus, *Poeticon astronomicon* (Venice: Erhard Ratdolt, 1482). Reproduced by permission of Linda Hall Library of Science, Engineering & Technology, Kansas City, Missouri.

For Nita Krevans

Contents

Introduction

This translation makes one of the most influential poems in antiquity accessible to interested readers and will, I hope, draw attention to an important but underappreciated poet. Aratus' *Phaenomena* is a didactic poem, a practical manual in verse which teaches the reader to identify the constellations and predict changes of weather. Though poetry and the sciences are today often regarded as opposing modes of expression, they fruitfully coincided in classical literature, and Aratus' *Phaenomena* argues for their compatibility. Like Aratus' original, my translation endeavors not only to teach the constellations and weather signs but also to help the reader understand why this knowledge is important. The following introduction provides historical and literary context and traces the reception of the *Phaenomena* in antiquity.

History and Biography

When Alexander the Great died in Babylon in 323 BCE, the empire he had carved out of mainland Greece, the Eastern Mediterranean, and the Near and Far East rapidly fragmented into kingdoms that were ruled by his former officers, known as the *Diadochi,* or "Successors." Aratus lived during the subsequent period of political instability and almost constant war, roughly from 310 to 240 BCE. Evidence for Aratus' life comes from four *Vitae* (*Lives*) preserved in manuscripts of Aratus and a fifth *Vita* in the late lexicon the *Suda* (*The Fortress*). In general outline, the *Lives* agree and most likely descend from a common source.[1] Thus, all the sources concur that Aratus was born at Soli in Cilicia (on the southern border of modern Turkey, eleven kilometers west of present-day Mersina), and scholars have dated his birth circa 310 BCE. In 276 BCE Aratus was invited to Pella to join the court of the Macedonian king Antigonus II Gonatas ("Knock-Knees"), who was the grandson of Antigonus Monopthalamos ("The One-Eyed"), the most powerful of the Diadochi. In the previous year Antigonus Gonatas had ambushed and defeated an army of Gauls at Lysimachia on the Thracian Chersonese and, as a result of his victory, had claimed the Macedonian throne. In honor of this victory Aratus composed a *Hymn to Pan* which alluded to the panic that befell the Gauls during the battle. This poem has not come down to us. At the Macedonian court Aratus also composed his most famous poem, the *Phaenomena* ("Appearances"). After spending time at the court of Antiochus I Soter of Syria and completing an edition of

Homer's *Odyssey*, he returned to Pella, where he remained until his death, sometime before 239 BCE.

Determining Aratus' influences is difficult because of the paucity and nature of the evidence. Ancient biographers tend to make perceived literary influences into actual teachers and acquaintances.[2] *Lives* II, III, and IV attest that Aratus associated with the literary luminaries Philitas and Callimachus. Philitas of Cos (born ca. 340 BCE), a narrative poet and epigrammatist about whom we know very little, is generally assumed to be an important forerunner of the Hellenistic poets. There is no evidence that Philitas and Aratus ever met, and the biographers most likely assumed their acquaintance because the *Phaenomena* displays the metrical and esthetic refinements which Philitas is held to have promoted. Callimachus (310/305–240 BCE) is perhaps the most influential poet of Aratus' generation and is regarded as having established esthetic principles for the Hellenistic age. Though again we cannot be certain that Aratus and Callimachus met in person, there is clear evidence of reciprocal influence and a familiarity with each other's work, as will be shown below.[3] The *Suda* also attests that Aratus was a pupil of the poet Menecrates of Ephesus (born ca. 340 BCE), who composed a didactic poem in the style of Hesiod.[4] It is striking that this poem treated, among other topics, astronomy and apiculture, both of which appear, on a large or a small scale, in the *Phaenomena*.[5]

We are told that during the late 280s and/or early 270s BCE Aratus resided in Athens, where, if he did not meet the philosopher Zeno of Citium himself, he at least became acquainted with his teachings.[6] Zeno (333–264 BCE) gave lectures in the Stoa Poikile ("Painted Colonnade") and elsewhere in Athens throughout the early third century BCE, and his teachings eventually led to the formal "Stoic" school of philosophy. This tradition, however, may mean nothing more than that the biographers recognized Stoic cosmology in the *Phaenomena* and attributed the influence to an actual acquaintance with Zeno or certain of his pupils.[7] There are resemblances between parts of the *Phaenomena* and the contemporary *Hymn to Zeus* by Cleanthes (331–232 BCE), the successor of Zeno as head of the Stoic school. We are also told that Aratus was a student in Athens of Menedemus (ca. 350–ca. 277 BCE), founder of the short-lived Eretrian school of philosophy, and that he had some interaction, mostly likely at Pella, with the skeptic philosopher Timon of Phlius (ca. 320–230 BCE).[8] Diognes Laertius preserves an anecdote in which Timon advises Aratus on the emendation of Homer's epics,[9] but we have no evidence to corroborate Aratus' association with Menedemus. In this alleged network of literary and philosophical acquaintances, we can identify three strains—Hesiodic epic, Stoic philosophy, and Hellenistic poetics—which come together in Aratus' masterpiece.

The *Phaenomena* is divided into two parts. The first part (1–783), an adaptation of a mostly lost prose treatise by Eudoxus of Cnidus, instructs the reader in the constellations of the northern and southern skies and their risings and settings.[10] The second part (784–1189), sometimes referred to separately as the *Diosemeia* (*Weather Signs*), largely concerns the signs by which the weather can be predicted. It draws primarily from a work on weather signs referred to as *De Signis* (*Concerning Weather-Signs*) and attributed to Aristotle's pupil Theophrastus.

Influences on the *Phaenomena*: Hesiod, Early Science, and Early Stoicism

Hesiod

The *Phaenomena* is a didactic epic—that is, a poem composed in dactylic hexameter (the same meter as Homer's *Iliad* and *Odyssey*) that provides instruction on certain specific subjects and, more generally, on prosperity in daily life. The most important predecessor to the *Phaenomena* in the genre is Hesiod's early-seventh-century BCE *Works and Days*, and Aratus' emphasis on agriculture in the proem serves as a nod to this poem as its primary model. Unlike Hesiod, however, Aratus approached his didactic epic as a literary rather than an oral-performative enterprise. As Fantuzzi and Hunter explain, "Hellenistic *didaxis* is, at base, the interpretation of prior texts."[11] Whereas the *Works and Days* consists of folk wisdom handed down orally, Aratus synthesizes three texts: Eudoxus of Cnidus' treatise on astronomy, Pseudo-Theophrastus' *Concerning Weather-Signs*, and Hesiod's *Works and Days*.

The speaker of a didactic poem addresses the listener or reader in the second-person singular, like a lecturer or schoolmaster addressing a pupil. In the *Works and Days* Hesiod preaches to his good-for-nothing brother Perses, focusing on farming and sailing as representative of all human activities. Though Aratus' injunctions do at times suggest that he is addressing a farmer or an overseas trader or sailor, his addressee has no fixed career or interests. Fantuzzi and Hunter observe that this "colorless, second-person addressee, whom every reader will interpret personally, conveys the universality of Aratus' message."[12] Furthermore, whereas Hesiod stresses competition and divisions between kings and ordinary people, Aratus elides these differences, emphasizing that all humankind hopes for prosperity:

> Though every man must choose his own career
> And hope that he can get ahead, we all
> Cling to whatever omens rise or fall
> And trust our lives to momentary signs. (1137–40)

The *Phaenomena* presents the experiences of all people as comparable—"a consistent picture of universal needs in the face of the same problems and opportunities."[13]

Whereas Hesiod introduces himself as an authority, Aratus does not name or call attention to himself as an individual.[14] In fact, he does not even distinguish himself as an authority from the common run of humanity. Cleanthes' *Hymn to Zeus*, by contrast, is elitist: writing of the state of humankind, Cleanthes implicitly sets himself and other philosophers apart from the "bad" (*kakoi*) men who neglect Zeus' universal law and only pursue worldly possessions.[15] Aratus differs from the rest of humanity only in that he serves as a mouthpiece for the Muses (poetic inspiration and literary tradition) (16–18). He most likely deemphasizes his individuality because, as Hunter suggests, drawing attention to himself "would ill suit the Stoicising stress on the centrality of the fixed order of nature in which no individual is particularly important."[16] It is appropriate, then, that neither the speaker nor the addressee in the *Phaenomena* has specific characteristics.

Though the *Phaenomena* is, on the whole, more tightly organized than the *Works and Days*, Aratus creates the impression that he is following Hesiod's looser "associative" logic. I have found it helpful to imagine Aratus as a lecturer who has an outline of a presentation—first he must run through all of the northern constellations, then the southern, then describe the Milky Way, and so on. Within this framework, however, he feels free to digress, telling relevant myths, for example, and giving advice. Thus, as he moves through this outline, "the structure of the whole, in which the repetition of words, rather than neatly signposted transitions, act as a unifying force, is part of Aratus' *mimēsis* of the Archaic manner of Hesiod's *Works and Days*."[17] In short, the *Phaenomena* proceeds systematically from beginning to end, but Aratus strives to create the impression that he is rambling on.

The two poems also differ in their portrayals of Zeus and the nature of the world in which humans live. In the *Works and Days* Zeus is omnipotent and omniscient (267–69), and Hesiod cites numerous examples to prove that mortals and even gods cannot outsmart his fickle and inscrutable mind (*nous*) (483–84). Hesiod's Zeus is, at times, misanthropic, punishing humankind for whatever advantages it has happened upon. Early in the *Works and Days*, for example, he sends Pandora with the jar containing all evils into the world as punishment for men and retribution for Prometheus' theft of fire.[18] This dark side of Zeus hardly appears in the first three-quarters of the *Phaenomena*. Rather, we learn that he has organized the world as a *kosmos* ("orderly system"), at least partly for the sake of humankind (9–12). Accounting for the contrasting character of the god between the seventh and third centuries BCE, Kidd explains that "with the growth of science the natural world is better understood, and Zeus is now a helpful rather than a hostile force."[19] As organizing principle, he has made himself immanent in the recurring signs by which mortals can "make sense" out of the world and predict the future (821–26).

Though Zeus does figure as a hostile storm god in "Weather Signs," the poem as a whole is organized around the assumption that mortals can better their lives by learning to predict the future from signs that were established, at least partly for their sake, by the creator of the kosmos. Aratus' role is to serve as an intermediary between the divine order and his audience. In accordance with tradition, he calls on Zeus and the Muses to assist him:

> By your grace,
> I aim to rise as high as mortal may,
> Hymning the heavens while you light the way. (16–18)

Though at first sight the inspired knowledge for which he prays seems similar to Hesiod's, his approach to his source material suggests that he is not making an extravagant claim to special knowledge.

According to the conventions of didactic epic, the Muses inspire the poet to speak truthfully about subjects in which he has no personal expertise. After warning Hesiod that "they know how to tell many lies that sound like truth" but also "know how to sing reality, when [they] wish" the Muses proceed to fill him with the ability to sing of the past and the future.[20] Though he concedes that all he knows of sailing comes from once being a passenger, Hesiod feels confident that he can teach the aspiring merchant "the measure of the echoing sea" because "the Muses have taught him to make song without limit."[21] In the fifth and fourth centuries BCE prose writing developed as a rival medium for instruction, and "specialists" in various trades and pursuits challenged the didactic poet's authority to "teach" all subjects. Plato attempted to drive a wedge once and for all between poetry and special knowledge, making a distinction between *hoi sophoi* (the learned men) and *hoi poētai* (the poets). Given the rise of these rivals to poetic authority, it is only natural that Aratus, in contrast to Hesiod, makes no claim to unlimited knowledge. He escapes from accusations of teaching "lies that sound like truth" because he describes appearances (*ta phainomena*) that are visible to everyone and because he has based his poem on the expert written account of Eudoxus. As Hunter notes, Aratus presented himself as "an 'expert' or 'professional' poet, and part of his expertise lay in knowing where to find things out; like Callimachus, he can still appeal to the Muses (vv. 16–18), but the Muses were now to be found in libraries."[22] Aratus is most distinct from Hesiod in that he consciously takes Hesiod's poetry as a model and relies almost exclusively on literature for his material.

Ancient Astronomy

A number of references to prominent constellations appear in Homer's *Iliad* and *Odyssey*, and Hesiod preserves folk wisdom concerning the constellations as they pertain to the agricultural cycle. Thales of Miletus (ca. 624–ca. 546 BCE), we are told, is the first Greek thinker to have made a formal mathematical study of the heavenly bodies. Applying mathematical principles, he predicted an eclipse of the sun on May 28, 585 BCE, and divided the year into 365 days. Astronomy continued in the fifth and fourth centuries as a subject for sages and philosophers. Pythagoras (572–492 BCE), for example, speculated that the motions of the planets reflect ratios similar to those in musical harmony, and Anaxagoras (born ca. 500 BCE) argued that the moon is earthly in nature because it has ravines and plains and inferred that the moon shines with a "false" (reflected) light, probably on the basis of his observations of lunar eclipses.

The Athenians Meton and Euctemon (ca. 430 BCE) refined the work of Thales by devising a more rational method of reckoning the days by months according to the annual journey of the sun through the twelve constellations of the Zodiac. A variety of time spans with different numbers of intercalary months had been proposed to reconcile the Athenian twelve-month lunar calendar (354 days) with the solar year (365 days). Meton is attributed with proposing a nineteen-year cycle with seven intercalary months. The resulting lunar calendar, consisting of 235 months (6,940 days), almost exactly equals nineteen solar years. Aratus praises this cycle in the transitional lines which introduce Weather Signs:

> The nineteen cycles of the sun have earned
> Worldwide approval, so you should have learned
> The stars Night ravels from Orion's belt
> And wheels back to Orion and his svelte
> Retriever—both those setting in the tide
> And rising skyward. (805–10)

Meton and Euctemon incorporated the nineteen-year cycle into *parapēgmata*, which are inscribed stones listing risings and settings and other key dates.[23]

Eudoxus of Cnidus (410/408–355/347 BCE), a mathematician and astronomer, wrote a *Phaenomena* in prose (of which the first part of Aratus'

Phaenomena is an adaptation). Eudoxus, it is reported, was born on the island of Cnidus and first studied mathematics with Archytas (428–347 BCE) in Tarentum. We hear of him traveling to Athens around 387 BCE to study with the followers of Socrates. He allegedly became a pupil of Plato, with whom he studied for several months before, as the *Lives* claim, they had a falling-out. He is then said to have lived in Egypt for a time and to have pursued the study of astronomy and mathematics. We hear of him traveling from Egypt to Cyzicus and the Propontis and then south to the court of Maussolos (best known for the elaborate tomb or "Mausoleum" named after him). We are told that Eudoxus returned to Athens around 368 BCE with the students who had gathered around him during his travels. We are further informed that he eventually returned to his native Cnidus and there built an observatory and continued writing and lecturing until his death.

Eudoxus' contribution to Greek astronomy is considerable: he was the first to attempt a mathematical explanation of the planets' erratic movements and the first to introduce the astronomical globe. Those of Eudoxus' astronomical works whose names have survived include *Disappearances of the Sun* (concerning solar eclipses), *Oktaeteris* (on an eight-year lunisolar cycle of the calendar), *Phaenomena, Entropon* (on "cyclic" astronomy, probably based on Eudoxus' observations in Egypt and Cnidus), and *On Speeds* (an attempt to account for erratic planetary motions). Though none of these works survive intact, the astronomer Hipparchus (second century BCE) cites from Eudoxos' *Phaenomena* in his commentary on Aratus, and we are able to get some sense of the ways in which Aratus adapted his source material.

Scholars have argued that Aratus' *Phaenomena* is "a virtuoso literary exercise with no real 'didactic' purpose."[24] According to this theory, wordplay and the incompleteness of the instruction undercut the poem's stated objective—to teach the constellations and weather signs. Volk claims that "the main appeal of didactic poetry in the Hellenistic age lay in the very fact that poets attempted to tackle such unwieldy, unpoetic, technical subjects"; she herself concedes, however, that this conception better applies to Nicander's treatises on snakes and medicines (second century BCE), the *Theriaca* and *Alexipharmaka,* than to Aratus' *Phaenomena.*[25] To begin with, the constellations are not an "obscure" subject:[26] the poem's audience would have seen them nightly (and more brightly) in the ancient world. Nor are the constellations an "unpoetic" subject; they are embedded in the Greek poetic tradition, appearing frequently in Homer's *Iliad* and *Odyssey*, in Hesiod's *Works and Days,* and sometimes at length in later poetry (e.g., Euripides' *Ion,* 1144–58). Furthermore, Aratus takes pains to make technical concepts accessible and uses only rudimentary mathematics. One assumes that if Aratus had intended simply to make art for art's sake, he would not, for example, have so thoroughly described the northern and southern skies but would have picked and chosen those constellations which lent themselves to poetic play.

Volk also contends that the poem is merely a literary exercise because it purports "to give advice for situations in which no member of this urban society would ever find himself (thus, the *Phaenomena* is geared toward the needs of farmers and sailors)." She later notes, however, that, "despite some references to the student's potential interest in farming and sailing . . . his character remains so undefined that one might in many instances take the 'you' simply to be a generic 'one.' "[27] As an Everyman, hoping, like us all, to prosper by anticipating future events, the addressee would not have been exclusive of a learned urbanite.

I see no reason to doubt the sincerity of the poem's instruction. Poetry, like science, has its own conventions, and wordplay does not undercut didactic intent any more than if a professor were to make a pun during a lecture. The first half of the poem provides a thorough description of the constellations visible from the northern hemisphere. In the second half, representative weather signs are selected, but, as Fantuzzi and Hunter explain, " 'didactic' poetry does not have to be comprehensive to be 'didactic'."[28] Aratus, in fact, makes the same point himself: "why / Name all the signs of trouble in the sky?" (1074–75). Like a professor giving an "outreach" lecture, he gauges the register to what we would today call the general reader, altering Eudoxus' text with didactic aims in view. Hutchinson observes that "even the most conspicuous patterns in his verses serve an expressive end: they highlight the meaning and further the sense of articulate discourse."[29] In the end, I believe the *Phaenomena* is genuinely didactic because it works—it effectively teaches what it proposes to teach.[30]

Aratus' Use of His Sources

In order to show how the poem operates, it will be helpful to take a look at some of the ways in which Aratus adapted source material. First, since he intends his *Phaenomena* to be a practical manual rather than a scientific treatise, he shifts the point of view from Eudoxus' mathematical and objective perspective to that of the observer. In addition, whereas Eudoxus attempted to account for the erratic and at times retrograde motions of the planets with epicycles within cycles, Aratus emphatically states that he feels daunted before their complexity and will omit them to return to "fixed signs and consistent things":

You cannot find
Their courses relative to some fixed star

Because they wander, weave and veer too far
In their broad cycles through the skies . . .
I cower
Before erratic motion—give me power
To speak of fixed signs and consistent things. (476–83)

Aratus here employs a rhetorical device called *praeteritio*, which calls attention to the subject the poet will omit. Rather than simply leaving the planets out, Aratus announces the omission to focus attention on what is orderly about the universe. The planets, by the very nature of their erratic movements, do not help humankind predict changes of season.

Aratus frequently lightens technical descriptions with mythological asides and wordplay, and his tendency to speak of constellations as the gods and mortals whom they represent frequently leaves distinguished characters in awkward and surprising positions. For example, Queen Cassiopeia ends up taking a dive beneath the horizon:

Sad bride
Of Cepheus, Cassiopeia, trails her daughter
And like a tumbler flips into the water
Headfirst; chair, feet, and knees are all we see—
Punitive shame for slighting Panope
And Doris. (690–95)

Rather than smoothing over the awkwardness of this partial setting, Aratus encourages the reader to visualize it with a simile. He then drags out the joke by suggesting that this embarrassment in the heavens is continued punishment for her boast. No doubt this humorous disjunction of constellation and mythic character did not occur in Eudoxus' technical treatise. In addition to lightening the tone from time to time, these awkward moments seem to encourage a healthy skepticism of mythic catasterisms.

Early Stoicism

A number of scholars have argued that Aratus does not take astronomy as an end in itself but rather as a means to promote a Stoic view of divine providence and the cosmos.[31] This interpretation is reasonable in that perhaps the largest departure from Eudoxos is the inclusion of Stoic themes. In the *Phaenomena*, as in the Stoic Cleanthes' *Hymn to Zeus*, Zeus is a life-giving force immanent in and coextensive with the kosmos. He can still be said to serve his traditional Olympian role as "father of men" because mortals, and in fact all living things, are part of the kosmos he framed and sustains:

Zeus fills roads and markets, brims
Oceans and bays. By Zeus alone we live,
Born as his children, too. (2–4)

All mortals here have the same filial relationship to Zeus that previously only mythic heroes like Cepheus (181) and Perseus (259) could enjoy.

Most significantly for the *Phaenomena*, Zeus establishes the constellations in the sky as signs for men:

Grouping the stars, he fixed them in the skies
As clusters and took care to organize
The annual astral passages in clear
Rotations, so the crops thrive every year. (9–12)

He also remains immanent in them and in the subcelestial signs.

Since Zeus
Conceals some causes, we cannot deduce
His whole plan at a stroke. By his consent
Learning proceeds. Everywhere immanent
In entrails, birds, storms, stars, he helps our race
To help itself. (821–26)

The poem's emphasis on observation also reflects the Stoic focus on particulars ("all existing things are particulars").[32] Still, I maintain that the work serves first and foremost as a practical guide to constellations and weather signs. The rational and orderly Stoic system provides an effective framework for instruction and encourages the reader to trust in this orderliness rather than suffer the "vertigo" that the erratic planets elicit (481–83). The poem ends not with praise of the orderly cosmos but with general advice about observing signs and preparing for the future. Thus, even with its Stoic cosmology and emphasis on observation, the *Phaenomena* is not, as Fowler claims, "Stoicism clothed in Hesiodic garb,"[33] insomuch as the astronomy does not serve as a means to propounding a Stoic world view. Rather, the Stoic cosmology serves as a stable framework within which the speaker promotes trust in the regular and reliable cycles of the constellations. Whatever Aratus' philosophical beliefs, the Stoic elements in his poem serve to advance the poem's didactic ends.

Hellenistic Poetry and the *Phaenomena*

In the early third century BCE King Ptolemy II Philadelphus gathered prominent literati to his court in Alexandria, and they served there both as court poets and librarians at the Great Library. These Alexandrian poets composed a literary (rather than an oral) poetry that was erudite and highly conscious both of itself and of the poetry that had come before it. Though some scholars have regarded the Alexandrian poets and the poets of the Hellenistic period in general as ancient "modernists" who attempted to break with literary tradition, Fantuzzi and Hunter have recently stressed their interest in maintaining continuity with earlier poetry: "The use of and allusion to the poetry of the past was for ancient poets part of the tools of the trade, a mark of their professional *technē* [craft]."[34] Conscious of their daunting predecessors, the Hellenistic poets often sought new kinds of poetic expression through the fusion of the genres that had been handed down to them (epic, lyric, tragedy, epigram, etc.). Rather than resulting in a wild and avant-garde *mélange*, Hellenistic genre-blending consists of using traditional meters for new purposes: Theocritus, for example, uses dactylic hexameter (the meter of epic and hymn) for pastoral poetry, and Callimachus uses elegiac couplets for hymn. Beyond these extensions of range, we do have a number of "experiments," but they are usually described, even by their authors, as novelties that are not to be taken seriously.

Hellenistic poets go to great lengths to situate their work in respect to that of earlier poets. Alongside the traditional model of "divine inspiration" (usually from the Muses), a Hellenistic poet often introduces visits to and visitations from an illustrious predecessor who had taught him the trade and provided him with models. The influence of these predecessors, though it comes from texts, assumes an air of exaltation similar to inspiration from the Muses and other divinities. Though Aratus makes no explicit declaration of his Hesiodic model, he makes this relationship clear in the invocation by emphasizing agriculture (the fundamental concern of the *Works and Days*). Callimachus, moreover, explicitly identifies Hesiod as Aratus' model in an epigram.[35]

Among Hellenistic poets, *leptotēs*—which can be translated as "slenderness," "refinement," or "elegance"—appears repeatedly as a key esthetic virtue. In the opening fragment of Callimachus' *Aitia* ("Causes") Apollo appears to the poet and gives the following injunction: "Keep your animals fat and your Muse slender *(leptēn)*" (fr. 1.24). Following suit in his description of the horns of the new moon, Aratus embeds this esthetic in an acrostic hinging on the word *leptē* (preserved in this translation as "slender") (837–43). Callimachus, in turn, responds to this declaration of esthetic loyalties, using the same adjective, *leptai*, in his praise of Aratus' diction.[36] Gutzwiller explains that "Callimachus makes it clear that Aratus' words are refined (*leptai*) not only because they imitate Hesiod selectively, choosing only what is sweet, but also because they result from the hard work of the scholar."[37]

Aratus is not afraid to apply this esthetic to novel themes. Most striking in this way are his mathematical conceits, such as the ratio of 5 to 3 (517–20), the equinoxes (534–38), and an armillary sphere (548–58).[38] He describes the constellation Triangulum thus:

Under Andromeda another sign
Has been delimited: three sides define
Triangulum, a clear isosceles.
The short side can be found with greater ease
Because it has more stars and shines more brightly
Than the two long sides. (237–42)

Aratus also encourages the reader to compare the elegance of the kosmos to well-crafted works of art. For example, he concludes his account of the circles of the ecliptic, the celestial equator, and the tropics with the following admiration of the kosmos:

Athena's workinghands
Could find no better way to solder bands
Into a sphere proportionate in size
Or spin them like these circles in the skies. (553–56)

The expression "hands of Athena" is a Hellenistic metaphor for perfection in the arts and crafts.[39] The simile suggests the craftsmanship that went into framing the kosmos and, by implication, equates the kosmos not simply with metalwork but with the poem that describes it. In this way, Aratus asserts that his poem reflects the elegance of kosmic design.

Etiology, or a story that accounts for sources and origins, was especially popular during the Hellenistic period. Hesiod's *Works and Days*, which provides etiological explanations for such things as why life is difficult, was attractive as source material. In fact, what is perhaps the most representative poem of the age, Callimachus' *Aitia* (or "Causes"), presents an encyclopedia of etiological accounts. Aratus, in turn, accounts for the origin of constellations and explains the catasterisms of individual constellations (the Bears, Ariadne's Crown, and the Maiden). He also retells an etiology which Hesiod presents in his *Theogony*: the origin of the Hippocrene (or "Fountain of the Horse") (218–25).[40] The etiological mode is problematic because it puts forth a single first cause for a phenomenon that may well have come about as the result of a lengthy process. Furthermore, as Fantuzzi and Hunter explain, "the etiological mode of explanation . . . offers a world which is 'invented' and then remains without change."[41]

Though ascribing a single first cause or first inventor to a phenomenon may well have seemed naïve even in Aratus' day, etiological reasoning was well-established in poetic tradition. Aratus frequently operates in this mode, ascribing, for example, the origin of the constellations to a "first inventor":

> Some one of those no longer living found
> A way to lump stars generally and call
> A group one name. Since he could not name all
> The stars minutely nor consider each
> Because so many in their circuits reach
> All round the world and often are the same
> In size and brightness, he devised a frame
> For clustered stars and sealed shapes in a border,
> And thus the heavens were marshaled into order. (384–92)

According to this conception, a single individual separated out all the constellations, and they have remained unchanged every since. At times, however, Aratus hints at an awareness that little credence should be given to these etiologies by playfully presenting multiple and conflicting accounts of the same myths. He twice tells the story of Zeus' infancy on the island of Crete, at one time relating a story in which two bears (who become Ursa Major and Minor) suckled the infant (31–33) and at another adhering to the traditional account in which the goat Amalthea does so (162–66). Though it is not impossible that all three animals could have suckled Zeus, the placement of these competing stories in relative proximity (130 lines apart) calls the authority of both into question. Similarly, when presenting the parentage of the Maiden (94–96), Aratus offers the reader "multiple choice" options, perhaps partly because there were variants in her myth but mostly because the variants open a variety of interpretive possibilities. By presenting variants as equally valid, Aratus compromises them all but allows them to operate simultaneously.

Aratus has an encyclopedic knowledge of the traditional stories which we call mythology and, with his tendency to cite obscure variants, is entirely in keeping with his fellow Hellenistic poets. Let us return briefly to Aratus' mythological aside on the two bears. Here, he presents a previously unattested version of a myth which is a combination of two others: (1) that of Callisto, a maiden attendant on Artemis who, once she is ravished by Zeus and expelled from Artemis' company, turns into a bear and becomes the constellation Ursa Major; and (2) that of the goat Amalthea, who is said to have nursed the infant Zeus. Aratus, however, undermines his own story in the telling: the collocation of the phrase "if the story is true" and "in Crete" in the original Greek evokes Epimenides' famous dictum: "All Cretans are liars."[42] Aratus seems to be hinting that readers should not take his new myth in all seriousness,[43] and this suspicion is confirmed when he cites the traditional version of the myth 130 lines later (162–66).

He also shows the Hellenistic propensity toward rapid and surprising shifts in tone:

> These clusters set while, opposite, not dim
> And meager but with belt and shoulders bright,
> Orion, trusting in his broadsword's might,
> Seizes his station on the eastern rim
> Of the horizon, dragging up with him
> The whole River completely from the east. (612–17)

Aratus switches from a description of the mythic Orion to the situation of his constellation, and "the notion of one constellation leading up another below it here gives to the figure a conception so extreme in its heroism as to touch the absurd." Thus, we find "differing tones and levels, strangely and strikingly juxtaposed and combined."[44] Though with these surprising shifts Aratus is in keeping with his contemporaries (Callimachus and Apollonius of Rhodes, in particular), still, on the whole, these combinations are "less drastic and disconcerting" than we would expect in a poem which "sustains an almost continuous didactic exposition."[45]

The language of the *Phaenomena* is largely the *Kunstsprache* of Homer—an artificial patois which combines spellings and morphology from a variety of dialects. Aratus often alludes to the *Iliad* and *Odyssey* and frequently makes a show of his erudition by using words that are rare or *hapax* (occurring only once) in Homer. He playfully uses the same phrase to describe the Scorpion (the constellation Scorpio) chasing Orion across the heavens (683) that Homer uses to describe Achilles chasing Hector in book 22 of the *Iliad.*[46] Though Aratus does refine his handling of the hexameter to Alexandrian standards, he shows more of a propensity towards archaizing than his contemporaries.[47] For example, he experiments with the somewhat clunky but very emphatic four-word hexameters that occur in the works of Hesiod, using them on several occasions to punctuate vignettes within the larger narrative.

On occasion Aratus travesties Homeric language. He applies the conventional Homeric patronymic (that is, referring to a hero by his father's name) to animals. Thus, as Homer refers to Achilles as "Peleides," or "son of Peleus," in the *Iliad,* Aratus calls chickens "scions of the cock" (1005). For purposes of bathos, he also inverts the patronymic formula, referring to frogs in terms of their offspring as "fathers of tadpoles" (989). Many of the meteorological phenomena and animals that figure in the second section of the *Phaenomena* (783–1174) have a long history as *comparanda* in epic similes. In these instances Aratus is drawing upon epic precedents both to glorify animals and forces of nature and to suggest that we can find in nature microcosms which parallel human behavior:

> ... black-winged flocks
> That talk of roosting with full-throated cries
> (One can imagine they are happy: each voice
> Cuts through air like a clarion call. Some flitting
> Around the leafy canopy, some sitting
> Among the boughs, they all roost round the crown,
> Wings clapping out their love of settling down
> or coming home). (1041–48)

In other descriptions of nature, Aratus goes a step further: whereas Homeric similes compare, for example, gathering soldiers to swirling winds or swarming bees, Aratus' descriptions suggest a sympathetic connection among the natural phenomena themselves:

> A squall will come
> When honeybees forsake the flowers and hum
> Safely in cells of wax and honeycomb. (1066–68)

The *Phaenomena* and Its Latin Translations: A Literary Rite of Passage

The *Phaenomena* enjoys a more illustrious pedigree of Latin translators than any other Greek poem. The list includes the likes of Cicero, Ovid, the near-emperor Germanicus (or possibly the emperor Tiberius),[48] and the emperor Gordian I. The poem, in effect, serves as a "bridge between Greek and Latin poetic traditions."[49] For readers like ourselves, who live in a time when didactic poetry, even more than poetry itself, is out of fashion, the popularity of the *Phaenomena* in antiquity is difficult to fathom. Thus, it will be useful to provide a brief history of the poem's Latin translations.

The Latin translators of Aratus fall into two groups: those who translated the work in adolescence (Cicero, Ovid, Germanicus/Tiberius, and Gordian I) and those who translated the work later in life (Varro Atacinus and Avienus). The task of translating the *Phaenomena* began as a capstone to Cicero's education and soon became a literary rite of passage for ambitious young men. After translating the poem, Ovid went on to a literary career, Germanicus (or Tiberius) and Gordian I went on to political careers, and Cicero went on to both. Varro Atacinus, though he translated the poem in adulthood, does not stand far outside this tradition, since he must have undertaken the project not long after learning Greek, again perhaps as a kind of capstone to that pursuit.

One-upsmanship is ubiquitous in ancient literary practice, and a liberal approach to translation promoted an organic, maturing version of *Phaenomena* into which mythological material, astronomical corrections, and updated scientific theories were freely added. Avienus, the last Latin translator of Aratus, synthesized the work of his predecessors and produced a version that is over 700 lines longer than the original 1,154-line poem.

Cicero (106–43 BCE)

Cicero's *Aratea* are divided into two groups: the fragments (*Ph.* f. 1–23 and l. 1–482) which translate parts of lines 1–757 and the fragments (*Prog.* f. 1–9) which translate parts of lines 758–1141. In Cicero's *De Natura Deorum* (*On the Nature of the Gods*) Balbus, the author's spokesman, recites from memory passages from the *Aratea* that were written, we are told, by a "still adolescent" Cicero.[50] Cicero most likely completed his *Phaenomena* (the first part of the translation) circa 87 BCE during his school years. He had to have finished the *Prognostica* before 60 BCE when he sent word to Atticus that this poem along with some other works was being sent for inclusion in his literary record.[51]

In *Brutus* Cicero claims that during his education he would "read and write and take notes every day," and Shiler regards the *Aretea* as a result of this process, claiming that Cicero "wrote set pieces which were either orations outright in form or compositions designed to develop that faculty; while 'the taking of notes' may have referred to Greek books from which he made Latin excerpts."[52] There is a native propensity towards didacticism in Latin letters, beginning with the Elder Cato's *Concerning Farming* (*De Re Rustica*) and continuing down through the history of the language. Aratus' *Phaenomena* would have been attractive to Cicero at least partly because there was no functional astronomical manual in Latin. Though the didactic nature of the original would have made the *Aratea* an excellent practice piece for a young orator, Cicero did not in the end regard them as an exercise. He speaks of them in affectionate terms that he does not use for his other translations.[53]

Townend misrepresents Cicero's intention when he claims that the *Aratea* "are to be judged solely as translations, and rather as translations of the type normally found in the Loeb classics than those of Dryden or Pope."[54] The distinction between these two kinds of translations has to do, not with the quality of the work, but with authorial intent. By editorializing, expanding, and reconceiving the original work, Cicero was not merely producing a crib for studying the original but crafting an independent work of art in the Latin language. If Cicero had intended his *Aratea* merely as a crib, he would not have taken the time to hammer them into hexameters or passed them along to Atticus to be preserved for posterity. In fact, in a letter to Atticus, Cicero pokes fun at a precious and hellenizing spondaic line, suggesting that this early work was over- rather than under-wrought.[55] Whatever the quality of Cicero's verses, Plutarch claims that at least for a time Cicero had the reputation of being "not only the best orator but also the best poet among the Romans."[56] In the gap between the death of Lucilius in 103 BCE and the publication of Catullus and Lucretius in the 50s, only Cicero's poetry exerted significant influence on later literature.

P. Terentius Varro Atacinus (82–36 BCE)

Though little of certainty is known about this Varro, Jerome claims that he was born in 82 BCE and learned Greek at the age of thirty-five (47 BCE).[57] Thus, we have a *terminus post quem* for his adaptation of the *Phaenomena* into a work called the *Ephemeris* (*Weather Forecast*). The two extant fragments (both concerning weather signs) do not give us much to go on. In one passage Varro borrows from Cicero's *Prognostica* but modernizes the style by cutting out two of Cicero's cumbersome and archaic compound adjectives.[58] As we shall see, improvement upon Cicero's *Aratea* becomes a kind of group effort that spans the history of the Latin language. It is noteworthy that Varro, unlike any other of the Late Republican and Early Imperial translators of Aratus, seems to have given much of his mature creative life over to translation.

Ovid (43 BCE–17 CE)

Though the two extant fragments of Ovid's *Phaenomena* do not give much sense of the style of the work, Gee is "tempted to see Ovid's translation of Aratus as belonging to the poet's youth." Speaking of both Cicero's and Ovid's translations, she argues, "It is logical that translating Aratus' *Phaenomena* might have been a good way of cutting one's poetic teeth, since part of Roman education, which incorporated astronomy, could have been the memorization of this poem as a way of learning the constellations and their positions."[59]

If translating the *Phaenomena* was a kind of literary rite of passage, further explanation needs to be given for its popularity. Though Cicero's translation satisfied the need for a practical manual, it contained interpretative errors and exhibited a poetic technique that rapidly became dated. Thus, a whole series of Early Imperial translators and aspiring literati displayed their mastery of verse-craft by translating the *Phaenomena* and perhaps "beating Cicero." This rite of passage theory directly applies to Ovid. As Hinds explains, "The translation of the well-known Hellenistic poem is surely the sort of exercise that Ovid would have set himself while first trying out his powers, rather than in his maturity, when daring innovation in subject-matter was his rule."[60] At least some of Ovid's mature work, however, looks back towards his juvenalia, for he must have reworked much of his *Phaenomena* in his late work, the calendric *Fasti*.

Germanicus (15 BCE–19 CE) / Tiberius (42 BCE–37 CE)

A Latin translation of the *Phaenomena* has traditionally been ascribed to Germanicus, but Gain and Herbert-Brown have proposed Tiberius as an equally likely alternative.[61] Germanicus would have translated the *Phaenomena* at a young age (he died at the age of 33), and Tiberius would have had time for such an undertaking only before his early appointment to *quaestor* in 23 BCE. The translator (whichever it might have been) at times emends Aratus, drawing some corrections from Hipparchus' commentary. For instance, he assigns Ophiunchus' feet to different stars and inserts an etiological myth for the constellation Arctophylax. He echoes Cicero's and Ovid's versions and romanizes Aratus by removing Greek and inserting local customs.[62] In his version of the "Golden Age" passage (97–111), for example, the translator omits a Greek custom of giving judgment and introduces private ownership of land (a concept of which Aratus makes no mention).[63]

Gordian I (ca. 159–238 CE)

We do not possess a single scrap of Gordian's Aratus, yet he best fits the rite of passage model. As a young man, he composed and translated poetry. Even more significantly, he "wrote these works so that Cicero's poems might seem very much out of date."[64] The purpose clause is telling here: Gordian sets out to surpass Cicero by providing contemporary and improved versions of Cicero's poetic works. Yet Gordian, like Cicero, left poetry behind and went to the courts when he reached maturity: "He did these things as a boy. Later, after he grew up, he litigated in the Athanaeum, sometimes before the emperor."[65] Gordian's youthful interest in outdoing Cicero was not one man's ambition but the expression of long-standing tradition.

Avienus (fourth century CE)

Avienus wrote at least three didactic works: *Descriptio orbis terrae*, a geography based on a Greek original by Dionysius Periegetes; *Ora maritima*, a description of the western and southern coasts of Europe; and *Aratea: Phaenomena* and *Prognostica*. Our last Roman translator of Aratus, Avienus is also "the last representative of a series of cultured Romans who adapted

Greek science, in particular Greek didactic poetry."[66] His *Phaenomena* totals 1,878 lines (728 longer than the original). The increased length results from the inclusion of more mythological material and contemporary scientific theory. For example, Avienus identifies Aratus' mysterious "On-His-Knees" (63–67) as Hercules (169–193) and inserts the late-antique theory of planetary double movement (911–23) at a point where, in the original, Aratus despairs of giving an adequate account (475–83).

Though the date of Avienus' birth is not known, the *Aratea*, most likely his third work, must have been completed later in his life, sometime after 387 CE.[67] Like Varro Atacinus, he must have given most of his adult life over to translation and thus did not translate the *Phaenomena* as a rite of passage at the end of his education. Rather, his version is the most mature expression of a living, evolving translation that spanned six centuries. He made his contribution to what had become the group effort of some of the greatest men in Roman history.

France and the Caliphate: Medieval Translations

During the medieval period we find a continuation of the Latin translation tradition (though in a less literary form) and the first Arabic translation of Aratus. In the first half of the eighth century CE an anonymous translator in northern France made a word-for-word Latin equivalent of Aratus' original, and in the second half of the century a different hand made revisions to it.[68] This crib is now known as the *Aratus Latinus*. Its Latin is clumsy and often mere gibberish. Le Bourdellès notes that at this time Byzantine scholars rarely visited France, and monks acquired their knowledge of Greek piecemeal from grammars and dictionaries.[69]

The Persian polymath Al-Biruni (973–1048 CE) cites Arabic translations of the first ten lines of the proem and the Maiden episode (92–134) in his *India*, an encyclopedia of Indian religion and cosmology.[70] The Arabic translation from which these citations were drawn has not come down to us but is attributed to Abu't-Taiyib Tahir ibn al-Husain (775/6–822/3 CE). After serving as a general under Caliph Abu Jafar al-Ma'mun ibn Harun (786–833 CE), Tahir ibn al-Husain went on to become governor of Khurasan and founded the Tahirid dynasty. Honigman suspects, however, that Aba 'Uthman Sahl ibn Bisr, a Jewish astronomer who served at the court of Tahir ibn al-Husain, was more likely the translator.[71]

With the Copernican revolution in the early sixteenth century and the publication of thorough and illustrated star maps such as Johann Bayer's *Uranometria* (1603), the *Phaenomena* ceased to be regarded as an authoritative scientific text.

Translation Methodology

In the twentieth century, translators of the *Phaenomena* tended to focus almost exclusively on content, more or less neglecting the poem's formal elements. The two most frequently read English translations were intended as service translations or cribs for the Greek original. Thus, though Greekless readers had access to the content of the original, they were left without the poetry. Since form and content are inseparable, the translator who intends to capture both the word and the spirit of the original must find not only appropriate words for the content but an appropriate form to replicate that of the original. Since Aratus wrote metrical verse, a free verse translation would be, by its very nature, a misrepresentation. I wanted to translate the *Phaenomena* into a form that helps the reader capture the feeling of the poem and settled on rhyming couplets as the traditional vehicle for didactic poetry in English.

English literature in the seventeenth century and the first half of the eighteenth developed in a manner somewhat similar to Greek literature in the fifth and fourth centuries BCE: prose came to rival verse as a popular medium. As a result, the particular virtues of prose and verse became clearer by the contrast. Writing of the two media in his introduction to an anthology of the poets of the English Augustan Age, W. H. Auden explains: "Verse, . . . owing to its greater mnemonic power, is the superior medium for didactic instruction. Those who condemn didactic poetry can only do so because they condemn didacticism and must disapprove *a fortiori* of didactic prose. In verse, at least, as the Alka-Seltzer advertisements testify, the didactic message loses half its immodesty."[72] The rhyming pentameter couplet established itself as the dominant meter of the Augustan Age and was adapted to a variety of genres and registers. Given its tendency to sound epigrammatic (or even sententious), the rhyming couplet functioned especially well as a vehicle for didactic verse, and contemporary English-speakers still associate rhyming couplets with didactic proclamations. Take this couplet from Pope's *Essay on Man*:

> Know then thyself, presume not God to scan;
> The proper study of Mankind is Man. (II.1–2)

Furthermore, rhyming couplets preserve the integrity of the line within the larger work in much the same way as the meter of the original and readily accommodate and reinforce antithesis, parallelism, and other rhetorical structures (such as chiasmus).

Finally, a note on the editorial apparatus. In addition to line numbers for the translation, I have provided the corresponding line numbers from the Greek original; the latter appear in brackets every fifteen lines. Following the translation are two appendices—one explaining the constellation risings and settings that figure so largely in Aratus' poem and the other listing the names of the Latin constellations. Comprehensive notes on the *Phaenomena*, keyed to the line numbers of the English translation, follow the appendices.

Translating the *Phaenomena* has been a capstone to my education, as it was for so many noble translators before me (Roman or otherwise), and it has been an honor to follow in their footsteps.

Notes

1. The most likely author of the original source for these *Lives* is the Stoic Boethius of Sidon (ca. 150 BCE). Cf. Scholion on Aratus 1091; Cicero *On Divination* i.8.13.

2. Lefkowitz 1981, 137: "The poverty of information preserved about the Hellenistic poets points up how little the Lives tells us about what we most want to know: why poets wrote and how they worked. The notions of discipleship and emulation only bear witness to the most obvious correspondences: nothing is said about intellectual influence, training or what they read."

3. Callimachus in *Palatine Anthology* 9.507 = 27 Pfeiffer = 56 Gow and Page; Prax. fr. 16 Wehrli.

4. Varro *On Farming* (*De Re Rustica*) i.1.9, iii.16.18; Pliny *Natural History* Ind. 8 and 9, 9.17; *Etymologicum Magnum* s.v. "ethmos."

5. Scholion on Pseudo-Euripides' *Rhesus* 529; Varro *On Farming* (*De Re Rustica*) iii.16.18; Pliny *Natural History* Ind. 9 and 9.17. For Aratus' bees, see lines 1076–78.

6. *Vita* III. Hunter (1995) raises some doubts about Stoic influences on Aratus and suggests the poet's alleged association with Zeno and his followers in the *Lives* may be a retrojection of "a Stoicising interpretation." Nonetheless, he concedes that interpretation of the "Stoic" elements in the *Phaenomena* is "to some extent unavoidable."

7. *Life* IV claims an acquaintance with the Stoic philosopher Persaeus of Citium (306–243 BCE).

8. *Suda* s.v. "Aratus."

9. Diogenes Laertius, *l.c.* 113; cf. *l.c.* 110.

10. *Life* III states that the treatise which Aratus adapted was entitled the *Phaenomena*; *Life* I claims that it was called *Katoptron* (*The Mirror*).

11. Fantuzzi and Hunter 2004, 240.

12. Fantuzzi and Hunter 2004, 231. Cf. Bing 1993; Fakas 2001, 94–100.

13. Fantuzzi and Hunter 2004, 238.

14. He does, however, introduce a pun on his name in the second line of the poem which serves as a sort of signature, deferentially placed after the name of Zeus. See Levitan 1979, 68, n. 18; Kidd 1981, 355; Bing 1990, 191–95.

15. Cleanthes *Hymn to Zeus* 20–31.

16. Hunter 1995.

17. Fantuzzi and Hunter 2004, 224–25. Cf. Fakas 2001.
18. Hesiod *Works and Days* 53–105.
19. Kidd 1997, 8.
20. Hesiod *Theogony* 26–28.
21. Hesiod *Works and Days* 646–62. Cf. Hunter 1995.
22. Hunter 1995.
23. For *parapēgmata*, see Appendix 1.
24. Fakas 2001, 5–38; Fantuzzi and Hunter 2004, 226.
25. Volk 2002, 54–55.
26. Volk 2002, 55.
27. Volk 2002, 55, 56.
28. Fantuzzi and Hunter 2004, 234.
29. Hutchinson 1988, 215.
30. The poem's efficacy is at least one of the reasons for its popularity among the Romans, as I discuss in the section on Latin translations in this introduction. I had the pleasure of reciting portions of my translation of the *Phaenomena* on March 13, 2000, at the Minnesota Planetarium in Minneapolis. An astronomer pointed out the stars and constellations with a laser pointer as I proceeded through Aratus' descriptions of them. Both the astronomer and the audience remarked that they were surprised by the poem's accuracy and effectiveness.
31. Effe 1977, 40–56; Fowler 1989, 161–63; Gee 2000, chap. 3.
32. Long 1974, 119, 141.
33. Fowler 1989, 163.
34. Fantuzzi and Hunter 2004, vii.
35. Callimachus in the *Palatine Anthology* 9. 507 = 27 Pfeiffer = 56 Gow and Page, 1.
36. Callimachus in the *Palatine Anthology* 9. 507 = 27 Pfeiffer = 56 Gow and Page.
37. Gutzwiller 2007, 33.
38. Kidd 1997, 26–28.
39. Kidd 1997, n. 529.
40. Hesiod *Theogony* 6.
41. Fantuzzi and Hunter 2004, 50.
42. Alluded to in Callimachus' *Hymn to Zeus* 1.8. A later formulation of this dictum—" 'All Cretans are liars,' the Cretan said"—is known in philosophical circles as the Liar's Paradox.
43. Kidd 1997, n. 30.
44. Hutchinson 1988, 219, 235. Hutchinson points out earlier that "Aratus emphasizes the splendor of the constellations both as purely visual phenomena and as animated figures" (217).
45. Hutchinson 1988, 236.
46. Kidd 1997, n. 646.
47. Kidd 1996, 33–36; Fantuzzi and Hunter 2004, 35–37.
48. Although this translation has traditionally been ascribed to Germanicus, Gain (1976, 16–20) and Herbert-Brown (1994, 176–77) argue that Tiberius is just as likely to have written it.

49. Fantuzzi and Hunter 2004, 224.

50. Cicero *On the Nature of the Gods* (*De Natura Deorum*) II.41.104

51. Cicero *Letters to Atticus* II.1. Although scholars have argued that the two sections were completed at different times, Courtney (1993, 149) dismisses this view.

52. Cicero *Brutus* 305; Shiler 1933, 19.

53. Cicero *Letters to Atticus* II.1.

54. Townend 1965, 117.

55. Cicero *Letters to Atticus* VII.2.

56. Plutarch *Cicero* 2.

57. Jerome *Chronicles* (*Chronica*) 82 BCE.

58. Courtney 1993, fr. 14.5–7. Cf. Cicero Prognostications (*Prognostica*) fr. 4

59. Gee 2000, 69 and n. 1.

60. Hinds 1987, 13.

61. Cf. Gain 1976, 16–20; Herbert-Brown 1994, 176–77.

62. Gain 1976, 83; Hinds 1987, 12–13.

63. Gain 1976, 86.

64. *Augustan Historians* (*Scriptores Historiae Augustae*) III.4.

65. *Augustan Historians* (*Scriptores Historiae Augustae*) III.4.

66. Herzog 1993, 368.

67. Jerome *Letter to Titus* (*Epistola ad Titum*) I.12.

68. Le Bourdellès 1985, 15, 69.

69. Le Bourdellès 1985, 149–50; Kidd 1997, 52–53.

70. Sachau 1910, I. 97, 383.

71. Honigmann 1950, 31.

72. Auden and Pearson 1953, 1.

PHAENOMENA

Invocation

Let Zeus be foremost—never may our hymns
Omit him. Zeus fills roads and markets, brims
Oceans and bays. By Zeus alone we live,
Born as his children, too. He deigns to give
Signs out of kindness to remind us rest 5
Must yield to work. He shows which soil is best
For cows, and which for hoes, and oversees
Seasons for sowing seeds and planting trees.
Grouping the stars, he fixed them in the skies
As clusters and took care to organize 10
The annual astral passages in clear
Rotations, so the crops thrive every year.
Our prayers begin and end, therefore, with Zeus.

Hail to a marvel, a helper we can use!
I greet you, Patriarch, your elder race, 15 [15]
And all you honeyed Muses. By your grace,
I aim to rise as high as mortal may,
Hymning the heavens while you light the way.

The Axle, the Poles, and the Northern Sky

The Axle and the Poles

Day in, day out, innumerable mixed
And scattered stars process above us. Fixed 20
Forever, never bending, an axle pins
Earth in the center of all; around it spins
Heaven on opposing poles, the axle's ends.
Though one cannot be seen, the other extends
Over the north.

The Bears and the Dragon

Two *Bears* surround this pole 25
(Which are at times called *Wagons* since they roll
Like wagon-wheels). Muzzle to behind,
They rear and dive with shoulder joints aligned
And bellies outward. If the tale is true,
Zeus the Almighty stellified these two 30 [31]
Because, near Ida, in his infancy,
They found him lying on Dicte's dittany

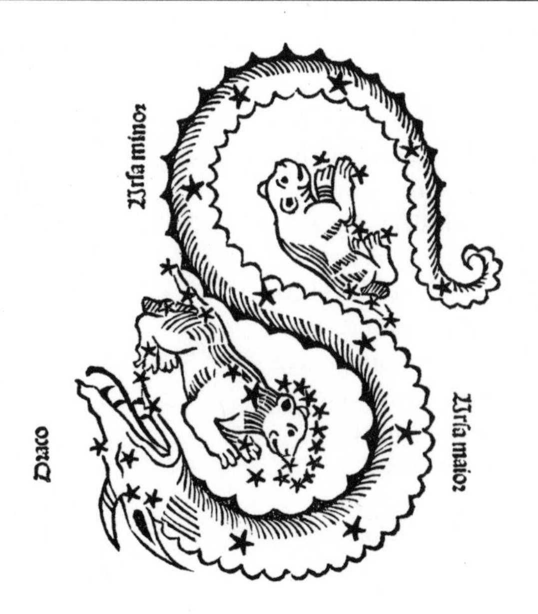

The Greater and Lesser Bears and the Dragon. From Hyginus, *Poeticon astronomicon*, D1 recto.

And picked him up and housed him in their den.
One year they nursed him while the elder men
Of Crete distracted Cronos from his son. 35

Men call one sister *Cynosure* and one
Helíke. The latter star-sign beacons Greek
Seafarers, while Phoenician sailors seek
Her sister cluster when they cross the sea.
Helíke scintillates more splendidly, 40
Obvious even at dusk, but Cynosure
More surely guides us (though she is obscure)
Because her group of stars rotates more tightly.
Setting their sights on her, Phoenicians rightly
Steer homeward.
Through the two Bears the breathtaking 45 [45]
Dragon meanders like a river, snaking
At great length far and wide. On either side
Of his extent, the Bears wheel with the tide
Beneath them. While the Dragon's nether tip
Teases Helíke's muzzle, his coils grip 50
The neck of Cynosure. Slipping around

Her head, the spirals loop her paw and bound
Back in a knot. A number of stars define
The Dragon's crest and cranial outline:
Two mark his temples; for eyes, another two; 55
And one point dots the monster's chin. Askew
Upon its neck, his head appears to nod
At Helíke's tail like an assenting god,
His mouth and brow as high as her tail's end.
His crest passes the point where settings blend 60 [61]
With risings in the north.

On-His-Knees, the Crown, the Serpent-Holder, the Serpent, the Scorpion, and the Claws

A silhouette
Wheels near it, squatting like a workman set
To some hard task. Since no one can divine
The sure form or vocation of this sign,
Men call him *On-His-Knees*. Again, bent over 65
On his knees, crouched like a heavy mover,
He swings each arm out from the shoulder joint.
He plants his right foot at about midpoint
Above the Dragon's head.
The god of wine
Made Ariadne's radiant Crown a sign 70
In heaven to hold her memory when she died.
It rolls behind the specter's back. Beside
His head's top stars you can discern the face
Of *Serpent-Holder*. Start from there and trace
The contours of his cluster. His shoulders shimmer 75 [76]
Even at mid-month moon; his arms (though dimmer
Because the stars fade on his left and right)
Still are conspicuous, if not so bright.
His hands grapple the *Snake* that binds his waist,
And his footsoles—relentless, firmly placed— 80
Tread monstrous *Scorpion*'s abdomen and eye.
The Serpent loops his arms in lassoes, high
And thick above the left but further down
On the right arm. Towards Ariadne's Crown
The Serpent's jaw inclines; now, under him, 85
Squint at the mighty *Claws*, for they are dim.

The Serpent-Holder and the Serpent. From Hyginus, *Poeticon astronomicon*, D7 recto.

The Plowman and the Maiden

Here comes *Arctophylax* the Wagoner
Behind Helíke. Since he seems to steer
The plow-like bear, men also call this sign
Plowman. Bright stars detail his whole outline, 90 [94]
And underneath his belt *Arcturus* wheels,
Distinctive from the rest.
Beneath his heels
A *Maiden* holds a golden ear of corn.
Whether, as poets rumor, she was born
The daughter of Astraeus (who, they say, 95
First sired the stars) or some god else, I pray
Her coming bring no evil. Some maintain
She used to walk earth and did not disdain
To meet the tribes of mortals face-to-face.
Though born divine, she joined the human race. 100
Her name was Justice then; packing the squares
And thoroughfares with seasoned counselors,
She promulgated what was fair and right.
Humans had never heard the hiss of spite,

The Maiden. From Hyginus, *Poeticon astronomicon*, E4 verso.

The bellow of quarrel, and the cry of war. 105 [109]
The wicked sea churned at a distance; oar
And sail had never shipped our livelihood.
Cows, plows, and Justice, giver of the good
And queen of peoples, furnished everything.

So long as land alone was nourishing 110
The Golden Race, she only lived on land.
Though later stooping low to hold the hand
Of the Silver children, she still walked the earth
Yearning for ways and men of greater worth.

From twilit foothills she would steal alone 115
And chasten humans in a harsher tone.
While gawkers hunkered on a mountainside
She would give speeches from the peak, deride
Their crimes and swear that she would keep her distance
However much they cried out for assistance: 120 [122]
"What trash your golden fathers have begotten!
O, your descendants shall be still more rotten—
Burdens of blood and war shall bow their backs,

Conscience shall crush them." She retraced her tracks
Down to the foothills when she had her say, 125
And all the people watched her walk away.
When they were dead, a fiercer Brazen race
Inherited—the first men to unbrace
Cows from a plowshare so that they might gorge
On flesh instead of grain, the first to forge 130
Marauders' trouble-making scimitars.
Justice turned misanthrope and joined the stars.
She still appears in heaven where at night
The Maiden wheels above us, near the bright
Plowman.
Along her right wing *Vintager* 135 [149]
Circles above her shoulders, and this star
Has just the same diameter and glare
As that beneath the tail of the Great Bear.
That Bear is marvelous, and marvelous
The stars that are nearby, before her paws. 140
Just glimpse those gleaming magnitudes, and you
Need not go squinting for a further clue—
Each of them can be easily divined:
There at her forelegs, there before her hind,
And there beneath the back of her rear knee. 145
Anonymous, they all wheel separately.

The Twins, the Crab, the Lion, the Charioteer, and the Bull

Under Helíke's muzzle *Twins* are placed;
The *Crab* scuttles along beneath her waist;
Beneath her hind paws a bright *Lion* rears.
The sun burns hottest here: just when the ears 150 [149]
Are stacked in sheaves, a stiff Etesian roars
Across the broad waves of the sea, and oars
Are out of season. May a broad-beamed frame
Convey me then, and may the helmsman aim
The prow aweather!
If you want to note 155
The *Charioteer*'s star-cluster, if the *Goat*
And Kids (who often see tars tempest-tossed
Upon the swell at night) have not been lost
Upon your ears, look leftward of the Twins

And find the great sign crouching as he spins. 160
Helíke's forehead also points the way.
The Holy Goat, who, as the legends say,
Once suckled Zeus, resides down from his collar.
This star is recognized by every scholar
As the Olenian Goat—it is a clear 165 [164]
Bright sign. Her Kids (the wrist of Charioteer)
Gleam faintly.
At his feet observe a *Bull*,
Hornéd and hunkering. So bright his skull,
One needs no other sign; so great the luster
All round his head, it beacons his whole cluster. 170
These vivid head-stars are by no means nameless
(In some circles they, in fact, are famous)—
Hyades is the bevy which highlights
The Bull's broad brow. A single star unites
Charioteer and Bull as yokemates, linking 175
Right foot to left horn-tip. Though, at their sinking,
The Bull goes down before the Charioteer,
Charioteer and Bull, when rising, rear
Their heads together.

Cepheus, Cassiopeia, and Andromeda

Cepheus' broken
And tragic race should never pass unspoken: 180 [180]
Zeus fathered them and saw fit to secure
Their fame in heaven. After Cynosure,
Cepheus himself stands upright and extends
His arms. The space from where the Bear's tail ends
To both his feet equals the gap from foot 185
To foot. Right near his belt you can make out
The first coil of the massive Dragon.
Hexed
Cassiopeia, his slender wife, wheels next;
At the full moon, her glimmering is frail.
Scanty and alternating stars detail 190
And decorate her form. Like bosses worked
Into a panel when the porter has jerked
The bolts back from the threshold's frame and thrown
Open the double-leaves, each on its own

Cassiopeia. From Hyginus, *Poeticon astronomicon*, D4 verso.

Her stars gleam thus. Her arms spread far and wide 195 [195]
From either shoulder. You would say she cried
Over Andromeda.
There, too, nearby,
Awesome *Andromeda* drifts through the sky
Vividly, under her mother. You can sight her
Without much squinting up and down, so bright are 200
Her toes, her sash, her shoulders, and her head.
Still she is strung up, still her arms are spread;
Even in the heavens chains and shackles have her.
She must hold out her pinioned wrists forever.

The Horse, the Ram, the Triangle, and the Fish

Over her head the massive *Stallion* rears 205
His chest. The top part of her figure shares
One star in common with his lower paunch.
Three other stars, conspicuous on his haunch
And shoulders, eke out corners for a square.
Although his head and neck are dim, the glare 210 [210]
Marking his muzzle almost could outshine
The whole quartet of pale stars that define

The Stallion. From Hyginus, *Poeticon astronomicon*, E1 recto.

His constellation. This unworldly Horse
On two legs only runs his cyclic course
Since at the navel he has been bisected. 215

It's fabled that this Stallion first directed
The gorgeous jet of fecund Hippocrene
Down from the heights of Helicon. Unseen
Pressure had built up in that sacred rock,
And when his forefoot kicked the peak, the shock 220
Released a river, and a watercourse
Formed on the mountain. "Fountain of the Horse,"
The shepherds named it then, or "Hippocrene."
You still can see stone spouting this pristine
Freshet among the men of Thespiae. 225 [223]
The Horse, however, spins through Zeus' sky—
There to be seen.

The *Ram* is capering
There also—swiftest, since he runs the longest ring
Around the skies but matches his rotation
With Cynosure's. Although his constellation 230
Is starless and jejune as if moonshine
Had blurred him, you need only trace a line

Out from Andromeda's girdle. Just a little
Beneath her he advances through the middle
Of broadest heaven, where Orion's waist 235
And the pinched tips of Scorpion are placed.

Under Andromeda another sign
Has been delimited: three sides define
Triangulum, a bright isosceles.
The short side can be found with greater ease 240 [236]
Because it has more stars and shines more brightly
Than the two long sides—this shape is slightly
South of the Ram.
Still further in the mouth
Of the expansive passage to the south
Two *Fish* are swimming. The bright one dives less deep 245
And can more clearly hear the sounding sweep
Of northern winds. Twin cables running through
Their distant tailfins unify the two
In one star-sign. The bright star on the tether
(The one that binds the cable-ends together) 250
Is called the Knot of Heaven. The nearest guide
To the north Fish is on the left-hand side
Of Andromeda, on her shoulder.

Perseus and the Pleiades

Forever over
The shoulders of staunch *Perseus*, her lover,
Her two feet circle. As he marches forth 255 [250]
Taller than all the figures in the north,
His gallant right hand gestures to the seat
Of his love's mother. Staring at his feet,
He walks his father Zeus' property.

The *Pleiades* cluster wheels near his left knee. 260
These sisters occupy small space in heaven,
And each sheds little light. The bards count seven
But only six show plainly to the eye.
Though no star slipped unmentioned from the sky
Since sages started handing wisdom down, 265
Still, I shall name all seven: Halcyon,
Sterope, Taÿgete, Kelaino,
Merope, Maia, and Electra. Though

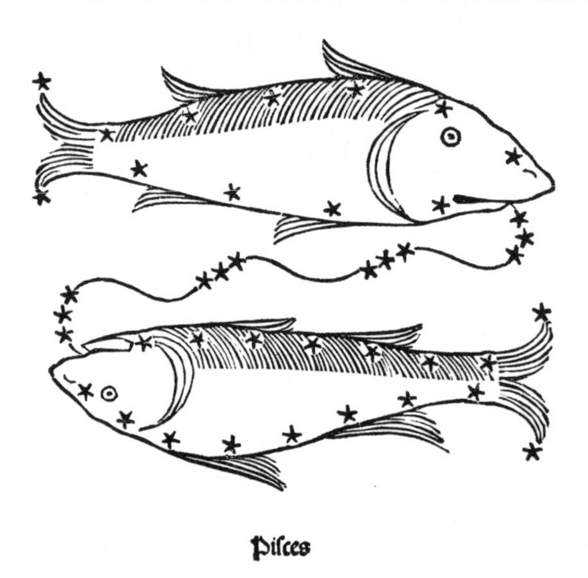

The Fish. From Hyginus, *Poeticon astronomicon*, E7 recto.

Obscure and small, their fame is broad: at dawn
And dusk they mark the year. Zeus drives them on 270 [265]
To beacon summer, winter, and the fertile
Season to break out plows.

The Lyre, the Swan, the Water-Bearer, Capricorn

A tiny turtle
Comes next, the one that Hermes perforated,
Scooped out in his crib and designated
The *Lyre*. He stellified the instrument 275
In front of the untitled specter. Bent
Near it, his left knee marks a side; the Swan
Uses its head to mark the cluster on
The other side.

Yes, under Zeus' sway
A mottled *Swan* goes winging on its way. 280
In patches here and there his stars are blurred;
Others are middling vivid. Like a bird
Who revels in fair-weather flight, he rushes
Westward when winds are calm. His right wing brushes
King Cepheus' right hand; the other wing 285 [280]
Flaps near the Horse suspended in mid-spring.

The Swan. From Hyginus, *Poeticon astronomicon*, C3 verso.

Where the Horse rears, the Fishes bob about
And *Water-Bearer* holds his right hand out
As if to touch its muzzle. He is borne
Into the heavens east of *Capricorn*, 290
Who runs before him on the lowest track
Where the sun pauses before circling back.

I pray that in this month you travelers never
Brave open water, pray that no waves shiver
Your spinning hull. If you embark at dawn, 295
You won't get far before the light is gone.
Days will be curt, and suns will never rise
The sooner for your shrill, benighted cries.
Winds from the south whip up the deep and smite
Hard when the sun and Capricorn unite; 300 [293]
From Zeus the frost falls and benumbs the sailor.
But all's one—surges churn beneath the tiller
All winter long. Hunkered like shearwaters,
We lie low in our vessel as it totters
And squint all round the ocean for some beach, 305
Some far-off sandbreak where the rollers reach
Land. A little brittle planking staves
Death off.

The Archer, the Arrow, the Eagle, and the Dolphin

When first you suffer higher waves,
Even a full month prior (when the sun
Sears *Bow* and *Archer*), doubt the night and run 310
Your prow ashore. The Scorpion appearing
Near sunup signifies the year is nearing
This violent month. The Archer draws his string
To launch a shaft against its threatening sting.
The Scorpion rises first but, hard upon, 315 [307]
The Archer follows him. Just then, at dawn,
Cynosure's head is at its greatest height,
Orion wholly has retired from sight,
And Cepheus gone down from hand to waist.

Aloft, another *Arrow* has been placed. 320
Somehow without a bow it was shot forth.
The Swan soars near the Arrow, to the north.
Another bird has taken flight nearby;
Small as he is, he swoops from sea to sky
Tempestuously when a new day starts. 325
His name: the *Eagle*.
With misty middle parts
A tiny *Dolphin* courses through the tide
Above the Goat. In couples, side by side
Four gems adorn him.
All these clusters run
Between the North Pole and the wandering sun; 330 [320]
Between the wandering sun and the South Pole
Different stars and other clusters roll.

The Southern Sky

Orion, the Dog, and the Hare

Mighty *Orion* stretches slantwise under
The sundered Bull. Let no night-watcher wonder
Which stars shine brightest in the night. Spread high 335
And broad, his cluster dominates the sky.

There, underneath his sweeping back, a dim
But dappled Guard *Dog* watches over him,
Jumping up on his two hind legs. A strip
Of black obscures his belly. From the tip 340

Orion. From Hyginus, *Poeticon astronomicon*, F1 recto.

Of his snarled lips the hottest star of all
Tosses its *searing* rays. Therefore, men call
This seether *Sirius*. Though feeble trees
Try to outwit the heat with canopies,
They wither when the Dog and Sun join powers. 345 [333]
His fiery arrows burn through leafy bowers
Like nothing. Some he scorches through the bark;
Others he toughens. One should also mark
This cluster's setting. A less piercing glare
Marks off the Canine's legs and paws.
The *Hare* 350
Forever runs beneath Orion's feet.
Tracking this cluster, always in a heat,
Sirius rises once he leaves his den
And then sits watching when he sets again.

The Argo, the Monster, and the River

Next to the Canine's tail stern-first proceeds 355
The *Argo*. Proper naval movements leads

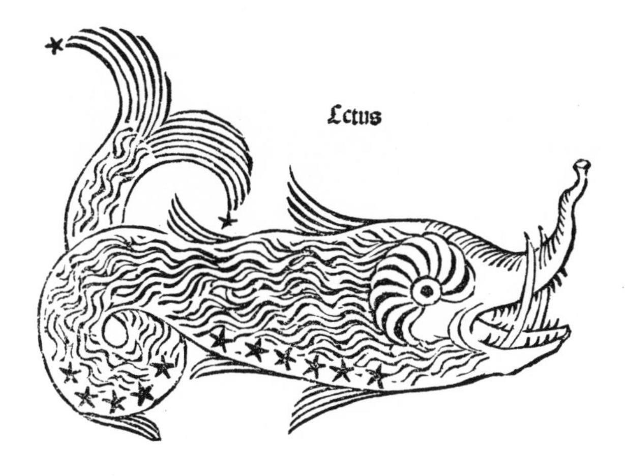

The Sea-Beast. From Hyginus, *Poeticon astronomicon*, E7 verso.

She not but backwards borne, as when tars turn
The stern of a real ship towards shore and churn
The shallows till they reach the beach rear-first—
So Jason's *Argo* cleaves the waves reversed. 360 [348]
Her aft decks all are radiantly spangled;
Her foredecks dark up to the mizzen. Dangled
Against the Dog's hind legs, the tiller whacks
His tail as he runs.
The *Monster* attacks
Nearby Andromeda with one great leap 365
Out of the south. As Thracian north winds sweep
Over her circling course, the south winds blow
The vicious Beast to her, down there below
The twin Fish and the Ram and just a bit
Above the starry stream. 370
Stars constitute
Weeping *Eridanus*' scant remains:
He trickles at the feet of gods and drains
Beneath Orion's heel. Tips of the tether
Which starts from either fish-tail come together
To link the Fishes in a single knot 375 [363]
Just at the Sea-Beast's nape—a single dot
Marking his topmost vertebra.

The Anonymous Stars

Lackluster
And miniscule, anonymous stars cluster
Under the gray Hare and between the Beast
And Argo's tiller. These cannot be pieced 380
Into the limbs of a prepared design—
Not like those star-signs marching in a line
Along set circuits as the years go round.

Some one of those no longer living found
A way to lump stars generally and call 385
A group one name. Since he could not name all
The stars minutely nor consider each
Because so many in their circuits reach
All round the world and often seem the same
In size and brightness, he devised a frame 390 [379]
For clustered stars and sealed shapes in a border,
And thus the heavens were marshaled into order.
No longer nameless, stars no longer rise
Into our sphere of vision a surprise.
Although most clusters have been named by us, 395
The hunted Hare treads these anonymous,
Dim stars.

The Southern Fish and the Water

There, under Capricorn, a lone
Fish bobs before the Monster, as if blown
Over the billows by a southern squall
Far from his brothers in the north. Men call 400
His cluster *Southern Fish.*
Dimmer and rarer
Stars are poured out beneath the Water-Bearer
And swirl through the bright heavens unrenowned
Between the Sea-Beast and this Fish. Around
The Water-Bearer's right hand, glistening 405 [392]
But feeble stars are coursing like a spring
That spills all over. Two of them appear
Much brighter than the others. Neither near
Nor far off from its partner, either wheels:
The first beneath the Water-Bearer's heels 410

Shines big and bright; the other sheds its luster
From underneath the Sea-beast's tail. This cluster
Is called the *Water*. Others, underneath
The forefeet of the Archer, like a wreath
Spin through the heavens.

The Altar

The Scorpion has poised 415
His fiery sting above an *Altar* raised
Down south. One sees it only briefly climb
Above the eastern margin, since the time
In which it can be seen runs counter to
Arcturus'. (Arcturus stays in view 420 [405]
A good long while above us, though the pressed
Altar swiftly disappears out west.)

Night is an old, old, crone who pities us.
She stuffs the Altar with conspicuous
Advice, and all because her greatest care 425
Is storm-beleaguered triremes. Everywhere
Around it telling signs of every sort
Guide sailors through rough seas and into port.

I beg you, pray that while you are at sea
The Altar at its zenith never be 430
Naked and radiant while layers of
Billowy clouds oppress it from above
(As happens in the fall when northerlies
Arise), because Night uses clouds like these,
In other seasons, for a harsh south wind. 435 [418]
Sailors are always foremost in her mind.
If they obey her early-warning signs,
Jettison cargo and adjust their lines,
Suddenly all their labors are more light.
But if, in contrast, sudden gale winds smite 440
Their galleys unprepared and blast athwart
Topsail and mainsail, they may find a port
Down in the murky deep. Still, if they pray
That Zeus the Storm-god turn the storm away,
If lightning strikes the north and all the men 445
Yarely perform their chores, they may again
Look on their mates alive. So, though the Altar

Brings stormy southerlies, they always falter
When lightning strikes the north. But when the star
That marks the Centaur's shoulder is as far 450 [431]
Westward as it is distant from the east,
When it is dim, and Night like a high priest
Sends forth distinctive signals from behind
Her sacrificial Shrine, no longer mind
The south but turn and face the winds that blow 455
Out of the east.

The Centaur, the Beast, Hydra, the Bowl, the Crow, and Procyon

Centaur appears below
Two other signs. His human face and chest
Rise up beneath the Scorpion; the rest
(The horse's back, the haunches, tail, and hooves)
Plods on beneath the Claws. His right hand moves 460
Interminably toward the rounded Shrine
But squeezes in its grip another sign—
The *Beast* (or so men titled it).

Behind
The Centaur, *Hydra's* sprawling spirals wind
And slither like a living snake's. His crest 465 [445]
Rears near the Crab; beneath the Lion's breast
His train trails backwards, while his long tail's tip
Menaces Centaur. His middle spirals grip
The *Mixing-Bowl*, and further down a *Crow*
Pecks at him. Procyon emits a glow 470
Nearby, beneath the Twins.

As seasons spool,
Study these signs as they return by rule
To where they started from. Each passing night
The stars are fixed forever in plain sight.

The Planets and the Great Circles

The Planets

Five other objects of a different kind 475
Swerve through the Zodiac. You cannot find
Their courses relative to some fixed star
Because they wander, weave, and veer too far
In their broad cycles through the skies. The time
When they will line up and renew the prime 480 [459]

Of Magnus Annus is far off. I cower
Before erratic motion—give me power
To speak of fixed signs and consistent things.

The Great Circles

Four circles gird the universe like rings.
Each serves as a much-needed reference-line 485
For self-completing seasons. Guide stars shine
Abundantly along their vast extent,
And these thin fitted circles complement
Each other: pairs of them, in breadth, are even.

The Milky Way

If under clear skies Night emblazones heaven 490
With all the brilliance she can give to us,
If there is no full moon to blunt the gloss
Of starlight, and if you have ever felt
Awestruck beneath a sky-dividing belt
Embossed with brightness while some friend or scholar 495 [476]
Is pointing out a span of creamy color—
You know the gorgeous Milky Way. In hue,
This belt outshines the other four. Though two
Match its diameter, the other two
Are smaller.

The Northern Tropic

Up where Boreas issues through 500
His gate, one of the small rings girds the sky:
Along it, both the Brothers' twin heads lie
As do the two knees of the Charioteer.
The left shoulder and leg of Perseus here
Swing through the heavens. Set above the bend 505
In Andromeda's right arm, beneath her hand
(Nearer the north), above her elbow (bent
Due south), its span transects the whole extent
Of her outreaching arm.

The Bird's neck moves
Along this ring behind the Horse's hooves 510 [487]
And Serpent-Holder's shoulders. Passing next
A bit north of the Maiden, it bisects
Lion and Crab successively: it cuts

Under the chest and downward to the guts
And privates of the Lion, then runs below 515
The Crab's hard shell and bifurcates his brow,
One eye on either side.
Divide this band
In eighths, and five out of the eight will stand
Above us in the sky; the other three
Loop through the regions that we cannot see. 520
The summer solstice sits here on the line
The Crab's divisions straddle and define.

The Southern Tropic

A circle in the south lies opposite.
Capricorn's middle portions rest on it.
The Water-Bearer's feet and Monster's tail 525 [502]
Come next. The Hare scampers along this trail,
And the Dog paws it just a little bit
With his hind feet. Next Argo crosses it,
Then Centaur's massive back, the Scorpion's sting,
And the resplendent Archer's bow. This ring 530
Marks out the low sun's southern terminus
Where he stops sinking and returns to us.
Cut it in eighths, and only three parts stay
In view.

The Celestial Equator

A ring broad as the Milky Way
Circles between these tropics and divides 535
The sky's extent into two equal sides.
When springtime gains or summer loses strength,
The day and night here find an equal length.
The Ram is carried lengthwise through this line.
The Bull's bright knees provide another sign. 540 [517]
Orion's belt lies on this boundary,
And Hydra entwines it. Though somewhat hard to see,
The Bowl stands and the Crow roosts here. The few
Stars that comprise the Claws reside here, too,
As do, in turn, the knees of Serpent-Holder. 545
Here, too, before the Horse's head and shoulder,
Flies Zeus' Eagle, herald of the weather.

The Ecliptic

An axis holds these parallels together
By forming a right angle. A fourth ring
Runs slantwise through the tropics, and they cling 550
Obliquely to it on opposing sides.
The more expansive central ring divides
It through the middle.
Athena's workinghands
Could find no better way to solder bands
Into a sphere proportionate in size 555 [529]
Or turn them like these circles in the skies—
The skewed ring holds the others as they loop
From dawn to night.

The Zodiac

Three of the rings come up
In parallel, then vanish. Each of them
Ascends at one point out on heaven's rim 560
And sets at one point on the other side.
The fourth ring spans as broad a swath of tide
As rolls from Capricorn's to the Crab's rising,
The hidden arc of this same ring comprising
As many settings on the other side. 565
If an observer's eye-beams could divide
Its length into six parts, two signs would fall
Into each section of the six. Men call
This ring the circle of the Zodiac.

The Crab and Lion travel on this track; 570 [545]
The Maiden next, the Claws, then Scorpion,
Archer, and Capricorn. Behind them run
The Water-Bearer, Fishes on their fins,
The Ram, the Bull thereafter, and the Twins.
Each year the sun must travel through them all, 575
Hauling in winter, spring, summer, and fall.
When half this circle cycles out of sight,
Half of it hoops across the sky at night.
As six-twelfths of it vanish, six appear.
Each night lasts for a single hemisphere. 580

It's smart to mark when these signs one by one
Rise out of the horizon, since the sun
Must come up somewhere in the Zodiac.
Weather permitting, it is best to track
These signs by their own stars. But overcast 585 [585]
Weather at night or heaps of mountains massed
On the horizon hide their nightly motion.
Therefore you need to find sure pointers. Ocean
On both its horns shows you the constellations
It wears like wreaths—these guide you to the stations 590
Where the twelve signs sit.

The Crab Rising

When Crab ascends,
Conspicuous stars rise from the world's ends
Out east or circle westward and go down
Beyond our ken: look for the setting Crown
Which still will be halfway above the line 595
Of darkness. Trace the Fish down to its spine.
Since On-His-Knees is spun head over heels,
His legs kick upwards while his torso wheels
From sight. The rising Crab drags Serpent-Holder
Down from his flexing knee up to his shoulder; 600 [578]
Up to its neck the Snake backs out of view.

The Plowman is no longer split in two;
His shrinking upper body joins his lower
Beyond sky's end. His setting measures four
Risings—four constellations balance one. 605
When he is tired of toiling in the sun,
He leaves his labor at so slow a pace
And takes so long unharnessing his brace
Of oxen that before his chores are done
Midnight is passed and darkness nearly gone. 610
Therefore these nights have been named after him.

These clusters set while, opposite, not dim
And meager but with belt and shoulders bright,
Orion, trusting in his broadsword's might,
Seizes his station on the eastern rim 615
Of the horizon, dragging up with him
The whole River completely from the east.

The Lion and the Maiden Rising

A rampant Lion lays those signs to rest
Which took a downward turn when the Crab rose.
Now dives the Eagle; On-His-Knees' left toes 620
And fingers stand alone above the verge.
The bright-eyed Hare and Hydra's head emerge
Out east, with Procyon and the Dog's bright feet.

Quite a few clusters swivel and retreat
Beneath the earth's edge when the Maid arrives— 625
The Lyre of Hermes sinks, the Dolphin dives,
And then the Arrow; from tail to west wing-tip
The Swan and then the lower River slip
Off into shadow; Stallion's bridle, nose,
And neck are lost as Hydra (who first rose 630 [602]
Beside the Crab) glides upward to the Bowl.
The Dog, his hind paws climbing from a hole,
Tows out the Argo stern-first with his tail.
Just when the Maiden clears the sea, its sail
Coasts half-seen through the night.

The Claws (Libra) Rising

You cannot miss 635
The rising of the Claws, dim as it is,
Because the Plowman and Arcturus climb
Completely into view at the same time.
All Argo now at long last leaves the ground.
Though Hydra's convolutions reach around 640
The heavens, still his tip is out of sight.
The Claw's ascent can only drag the right
Foot, shin, and knee of On-His-Knees in view.

This constant croucher, genuflecting to
A Lyre—what man, what marvel can he be? 645 [616]
We often watch him setting in one sea
And rising from the other in one night.
Still, only one of two legs is in sight
During the Claws' ascent. This sign (reversed
In that his head comes last and legs go first) 650
Spans, too, the rising of the Scorpion
And rising of the Archer. While the one

Carries his waist and torso back to us,
The other brings his head and left hand. Thus,
Odd On-His-Knees emerges upside down 655
In three successive stages.
Half the Crown
And the last taper of the Centaur's tail
Come into view just as the Claws first scale
The heavens. Following its head, the Horse
Goes down; tail-feathers follow the same course 660 [628]
Their Swan had flown. Andromeda's head sets.
The cloudy south wind wafts the Monster's threats
Against them all but Cepheus makes a stand
Up north and waves the Beast off with his hand.
Then, as it slithers backwards towards its den 665
And vanishes as far as its rear fin,
We start to lose his shoulder, head, and hand.

The Scorpion Rising

When Scorpion first rises from the sand,
The River meets the sea and empties there.
Only the galloping Scorpion can scare 670
Mighty Orion back to the abyss:

May you at last forgive him, Artemis!
They say Orion, after he had struck
The beasts of Chios dead, once dared to pluck
Even at the cloak no mortal hand should touch. 675 [638]
Orion went too far and did too much
For Oinopion. Stealing him a gift
From gods? The huntress goddess drove a rift
Straight through the island, and a Scorpion
Emerged and stung him for the wrong he'd done— 680
A monster massive as the mindful wrath
Of Artemis. So, when Scorpion finds a path
Into the sky, Orion takes his heels.

Do not ignore that Scorpion also wheels
The Monster and Andromeda outside 685
The visible night sky. And now the tide
Washes the head of Cepheus, and the west
Receives his dangling baldric (though the rest—
Feet, legs, and loins—will never reach the tide

Because the two Bears drag them back). Sad bride 690 [654]
Of Cepheus, Cassiopeia, trails her daughter
And like a tumbler flips into the water
Headfirst; chair, feet, and knees are all we see—
Punitive shame for slighting Panope
And Doris. As she meets the western verge 695
Of the night sky, a number of clusters surge
Out of the other ocean opposite:
More of the Crown arrives and brings with it
The Hydra's tail, which drags out of the east
The trunk and head of Centaur. Then the Beast 700
Surfaces in the Centaur's strong right hand.
But still the front hooves of the Man-Horse stand
Waiting the Bow. Bow lifts the Serpent's twists
And Serpent-Holder's arms up to the wrists.
Scorpion brings the Serpent-Holder's head 705 [668]
And Serpent's crest, its foremost coils still spread
Around his hands.

Given that On-His-Knees
Wheels upside-down, his lower extremities
First reach the heavens: legs and belt and breast.
Then come right shoulder, arm, and hand. The rest 710
(That is, his head and other hand) emerges
With the Archer.

The Archer, Capricorn, and the Water-Bearer Rising

When the Archer surges
Over the ocean, Hermes' Lyre as well
As Cepheus' stomach clear the swell.
Then all the Hunting Dog's effulgence, all 715
The hounded Hare, and all Orion fall.
The Kids and Sacred Goat now disappear
Gradually with the setting Charioteer.
They sit there shining on his massive arm
Distinctly, and they still can raise a storm 720 [682]
When sunrise crosses them. Some parts of him
(His head and waist and other upper limb)
Vanish as soon as Capricorn ascends.
His upper legs down to his furthest ends
Sink at the Drawer of the Bow's return. 725
Neither Perseus nor the spangled stern

Of Argo stay in heaven: his right knee
And foot alone poke from the western sea,
And Argo from the keel up to its rear
Deck founders. Even its forestays disappear 730
When Capricorn first leaps into the sky.
While Procyon sets, Bird, Eagle, and Arrow fly
Upward, as does the Altar, which is placed
Still further south.

When Water-Bearer's waist
Clears the horizon, the Horse's head and feet 735 [693]
Come coursing up. Out west a Night replete
With stars sends Centaur's tail from sight; and though
She cannot find sufficient space to stow
His shoulders, head, and breast-plate, she does pack
The Hydra's head-stars up with his coiled neck. 740
Much of the Hydra trails behind; but Night
Draws this, too, with the Centaur, out of sight
Just as the Fishes leap.

The Fish, the Ram, the Bull, and the Twins Rising

When they arise,
The greater portion of the Fish that lies
Beneath dark Capricorn forsakes the sea. 745
A little part, though, waits there patiently
Until the next sign rises. Piece by piece,
Tragic Andromeda's long forearms, knees,
And shoulders come in view, one side out front,
And one behind, just as the Fishes mount 750 [706]
Above the eastern sea. Her right-hand side
Follows the rising Fishes from the tide;
The left side rises with the Ram, in turn.

Then, with the Ram's ascent, you can discern
The Altar sinking low down in the west 755
And, out east, Perseus' head and chest.
Although some say the Ram, and some the Bull,
Brings up his belt, we can be sure his full
Form rises with the latter.

Fastened to
The Bull, the Charioteer is dragged in view 760
But only partially—the rest of him

Waits for the Twins to peek above the rim.
Before that, Goat and Kids and his left foot
Rise with the Bull, dragging the Monster out
Tail-first as far as its rear fin. The west 765 [721]
Then pulls the Plowman closer to his rest.
The Bull is one of the four signs that bring
The Plowman home to darkness—everything
Except his hand, that is, which stays up there
Circling the skies beneath the Greater Bear. 770

Take Serpent-Holder sinking in the brine
Up to his knees as a conclusive sign
That Twins have started rising in the east.
Cut off from both the oceans, the Sea-Beast
Completely rises, all his stars appear. 775
Now can a sailor, when the night is clear,
Observe the River trickle above the line—
But he awaits Orion, some sure sign
To teach when night will end or if his quay
Lies in the offing or still far away. 780 [731]

Indeed, the deities have given humans
Ubiquitous predictions, many omens.

Weather Signs

Proem

Don't you see? Moon declares a month is born
When in the west she grows a sliver of horn;
And when her beams are bright enough to splay 785
Night-shadows, she announces her fourth day.
Eight days upon her, she is half-concealed,
And only during mid-month fully is revealed.
So, as she grows or shrinks from phase to phase,
We measure out the passing of our days. 790

The twelve signs of the Zodiac portend,
Each in its turn, a sunrise at night's end;
And Zeus has further labored to make clear
The most auspicious times in the Great Year
To plow a fallow field or plant a tree. 795 [742]

Numerous sailors on an autumn sea
Safely admire storm clouds from afar

Once they have learned Arcturus or some star
Rising at dawn or dusk foretells a squall.
In fact, the telltale sun careens through all 800
Twelve astral figures as he brands the year.
Rising or setting now, he passes near
One marker or another, and the warning
Star where he starts each marks a different morning.

The nineteen cycles of the sun have earned 805
Worldwide approval, so you should have learned
The stars Night ravels from Orion's belt
And wheels back to Orion and his svelte
Retriever—both those setting in the tide
And rising skyward.
All these stars provide 810 [757]
Annual omens if you heed them well.
So learn them all. And if you tempt the swell
On shipboard, study signs that prophecy
Breakers at sea and storm clouds in the sky.

A little bit of watchfulness will reap 815
Boundless rewards, for the wise man will keep
Himself and others safer by foresight
When storms are brewing. On a windless night
Tars often trim their sails for fear of morning.
On the third dawn, or fifth, or without warning, 820
A storm will sweep down from the sky.
Since Zeus
Conceals some causes, we cannot deduce
His whole plan at a stroke. By his consent
Signals appear. Everywhere immanent
In entrails, birds, storms, stars, he helps our race 825 [772]
To help itself: the moon's expressive face
Will tell you she is halved or shrinking, growing
Or full. The sun sends signs when he is going
Down and every morning when he slips
In view. Of course, there will be further tips 830
To make your own concerning day and night.

The Moon

First, keep the two cusps of the moon in sight:
The twilight tints her rising and adorns

The third and fourth days with distinctive horns.
A study of their shape and coloring 835
Will teach you what a newborn month will bring:
Slender and lucid tusks in the third phase
Light windless nights, but if a crimson haze
Ensanguines horn, watch for a gale-force blast.
Nubblier horns in these two phases cast 840 [785]
Dull beams when storms or southerlies have blurred
Extensive swaths of air. If on the third
Rising, the twin horns neither dip nor dangle
But curve their tips upward at a right angle,
Expect the west wind. Should they fail to grow 845
By the fourth night, less friendly winds will blow;
And if the top horn seems a frowning mouth,
Beware of the north wind—but watch the south
When it leans backward.
When a perfect sphere
Rounds out her third phase, clouds will soon appear 850
Far off on the horizon; the more ruddy
She looks, the worse the coming thunder. Study
The moon each evening—crescent, tint, and quarter;
Watch as the splendor wholly fills her border
And then shades back again, begins anew. 855 [800]
You can foretell the weather from her hue.
A clear face means fair weather, while a blush
Informs the world that panting winds will rush
The earth. Remember, if dark blotches stain
A candid face, you should prepare for rain. 860

The third and fourth look to the seventh phase;
The seventh, the fourteenth. When half her days
Are spent, the monthly cycle is reversed—
The full moon eyes the waning twenty-first,
And so her cyclic prophecies descend 865
Back through the fourth and third days from the end.

A first, a second, even a third tier
Of halos sometimes circumscribe her sphere.
One halo means a calm or windy day—
Wind, if it breaks; calm, if it fades away. 870 [814]
Always expect foul weather from a double
Corona and be sure that far worse trouble

Follows tiaras. A still meaner squall
Howls when all three are dark, but—worst of all—
Disaster strikes whenever these are broken. 875
Just so, the changes of the moon betoken
Weather throughout the month.

The Sun

Do not neglect
Sun's daily journey westward; closely inspect
Sunrise and sunset for the surest signs.
May his eye open freshly with no lines 880
Or blotches when you hope for sunny skies.
No puffy clouds will mar the next sunrise
If he is glassy when cows go to graze
And amble home. If north- or southward rays
Splay from a sun with a bright core and hollow 885 [829]
Border, rain or wind will likely follow.

Stare straight into the sun's light, if you dare.
Check whether he blushes (often, here and there,
His surface reddens when a random cloud
Obstructs his beams) or whether a grainy shroud 890
Darkens him. Since a dark sun looks to showers
And blushes stir up winds, the allied powers
Of wind and rain strike when the two are mixed
Together. When his sloping beams are fixed
Upon one point or arching clouds beset 895
His rising, noon and evening, look for wet
Blustery weather. If his first rays poke
Blurrily through the evanescent smoke
Of morning, get inside before the storm.
Yet, if he spreads out and his sunbeams warm, 900 [846]
Melt and disperse the mist around him, clear
Weather will come. Clouds also disappear
After a steely dusk.

If rains have wet
Your fields by day, turn west to the sunset
And watch the clouds. If darkening clouds obscure 905
His shine and knock his beams astray, make sure
You seek out shelter. If the sun is rimmed
With ruddy clouds or utterly undimmed

Just as he dives and goes down out of sight—
You need not worry about showers that night 910
Or the day after. Batten down for rain
Whenever solar beams grow thin and strain
Dimly, as if the moon were in their way.
When late rays seem to straggle and delay
The dawn, or scattered crimson cloudbanks hover 915 [867]
Above the landscape, better run for cover.
When, before dawn, a hidden sun shines shadows,
Winds will arise or showers drench the meadows:
If he emerges in a thick dark curtain,
Rain is almost absolutely certain. 920
But if a slight murk dresses the sun's rays
In looser shadows, like a light mist-haze,
Expect the wind to rise.
Halos around
The sun mean trouble. The more tightly wound
And dark they are, the worse the storm will be— 925
Two halos forecast more ferocity.
Closely inspect the sun at dusk and dawn:
That dark cloud that we call "parhelion"—
Is it tinged crimson on the south or north
Or is it red on both? Zeus hurtles forth 930 [882]
Fierce storms when two low cloudlets sit astride
The sun. A lone cloud on the northern side
Means northerlies; a southern cloud foretells
Southerly breezes or rain-sodden squalls.
Study parhelia especially in the west. 935
If you fear rain, these omens work the best.

The Manger

The hazy Manger is another guide—
Look for it in the Crab. On either side,
Its north and south stars glow with feeble strength.
Mark them; they sit about a cubit's length 940
Apart. This pair of stars is called the Asses;
They keep watch on the Manger. Black cloud masses
Suddenly fill clear skies with stormy weather
If the two Asses slump too close together
And the Manger is all blotted out. 945 [901]
Again, when, without squinting, you can note

The Asses while the Manger is obscure,
Rain threatens. If the north Ass sheds a poor
And hazy glimmer while the other Ass
Vividly shimmers, southern winds will pass 950
Northward. Winds out of the north arise
Whenever the north Ass cuts through the skies
And the south Ass is dull.

Wind Signs

A tumid tide,
Beaches and breakers roaring far and wide,
Sea-caves echoing on a sunny day, 955
And mountains murmuring from far away—
These all are signs of wind. A shrieking heron,
Gliding in helter-skelter flight from barren
Ocean to sandy shoreline, always sails
Before the bluster of incoming gales. 960 [914]
And sometimes the storm petrels in fair weather
Tack as a flock and face fresh winds together.
When wild ducks and shearwaters beat the beaches
With restless wingtips as a cloud bank reaches
Still further out above a mountain's crown, 965
The wind will soon come up. The fluffy down
Of aging thistle long has been a sign
Of wind—its seed-tufts bobbing in the brine
In clingy bunches drifting here and there.

When lightning strikes in summertime, note where 970
The bright fork flashed—the wind will soon arise
From that same quarter. Stars shoot through the skies
Quite regularly; if their trails are bright,
Expect the wind to follow that same track.
When stars shoot forth and others fire back, 975 [930]
All kinds of wind could come from anywhere.
Then even our best weathermen despair
And no one knows where winds will rise or rest.

Rain Signs

When lightning strikes first east and south, then west
And north, a sailor fears the rising powers 980
Of waves beneath his planks and Zeus' showers

Upon his deck. These lightnings give a sign
Of rain to come. Salts also should divine
Showers from a rainbow with a double bar
Or a dark halo circling a star 985
Or clouds like fleece.
Birds of the bay and ocean
Commonly splash and stir up a commotion,
Or swallows flit around and around lakes
Or dive right in. Fathers of tadpoles (snakes
Love to devour these miserable things!) 990 [946]
Croak in the shallows while a tree frog sings
A throaty solo at the crack of dawn,
Or crows turn towards the breakers coming on
And cackle on the seashore, or perhaps
They dip their heads in, and salt water laps 995
Over their hackles or perhaps they leap
Into the waves, emerge, and get no sleep,
Pacing and gibbering along the shore.

The grazing cattle sniff the breeze before
A rain begins to fall; millipedes stick 1000
In clusters to a wall, and ants are quick
To haul their precious eggs out of their hole,
While worms, which once were called "earth's entrails," roll
And squirm helplessly on the ground. Take stock
In farm-raised fowl, the scions of the cock, 1005 [960]
Whenever they pick off their lice and cluck
And sound like pools of standing water struck
By constant dripping. Look for wind when flocks
Of daws and tribes of ravens shriek like hawks.
Ravens will also imitate the sound 1010
Of heavy raindrops splashing on the ground
Or cackle twice before they ululate
And flap. Restlessly ducks and daws will wait
Beneath the eaves, and herons rush to sea
Shrieking. Observe these signs attentively 1015
And you will keep ahead of a rainfall.

Take note when lice bite more than usual
Or snuff clings to the wick on a moist night
Or if the flame has an increasing height
In winter-time or sudden cinders rise 1020 [980]

Like bubbles or the flame flares up and dies,
Or if the bright birds of the islands fly
In tight formations through a cloudless sky.
Do not ignore when excess cinders float
Over the tripod and the cooking pot 1025
Or when strange spots like seeds of millet glow
In ashes—mark these signs if you would know
When the next rain will fall.

Signs of Fair Weather

Whenever a haze
Obscures the foothills while the summit stays
Cloudless, expect no trouble in the sky. 1030
And if tight clouds like granite ledges lie
Low near the surface of the ocean, warm
Weather will follow. Always fear a storm
On sunny days but note: all storms blow over.
Mark when the Manger emerges from cloud cover 1035 [996]
(That's when the Crab returns it to plain sight):
It clears when squalls dissipate. Steady light
Of candles and the owl's least jarring call,
Crows whistling imitations at nightfall,
A raven which at first caws twice, then squawks 1040
And squawks without a break, and black-winged flocks
Which talk of roosting with full-throated cries
(One can imagine they are happy: each voice
Cuts through air like a clarion call. Some flitting
Around the leafy canopy, some sitting 1045
Among the boughs, they all roost round the crown,
Wings clapping out their love of settling down
Or coming home), and cranes which flock together
Tending the same unswerving course—fair weather
Will follow all these omens faithfully. 1050 [1012]

Signs of Foul Weather

When all the stars are dim but you can see
No serried mass of clouds making a haze
Or any obstruction that could serve as cause—
No moonlight, but they just look faint—a storm
Is moving in. If cloud formations swarm 1055

Around a smother swirled up in the air
Or pass it drifting on their way, beware.

When geese with excess honking form a wedge
And rush to forage in your acreage,
Or when the crow (who lives nine lifetimes) caws 1060
At all hours of the night, when squawking daws
Keep talking late or chaffinch chat at dawn,
When seabirds seek land in an echelon
Or wrens and robins dive deep into burrows
Or flocks of daws at dusk exchange the furrows 1065 [1028]
For roosts, expect a storm.
A squall will come
When honeybees forsake the flowers and hum
Safely in cells of wax and honeycomb
Or when a flock of wandering cranes turns home,
Abandoning all travel. Never fear 1070
Wind-lifted gossamers on calm and clear
Mid-afternoons or candles flickering
Or no fire catching on the kindling—
These are not signs of storms to come.
But why
Name all the clues to trouble in the sky? 1075
For instance, signs of snow are found in clots
Of ash or in a ring of seed-like spots
Around a burning wick—and look for hail
When bright coal at its center wears a pale
Halo.

Signs of the Seasons

Observe the holm oak and its mast 1080 [1043]
And heed the dusky mastich. Crofters cast
Expectant glances all around the yard,
So summer never catches them off guard.
Oaks closely set with scanty acorns hint
At wicked winter storms. But branches bent 1085
With too much produce and well-watered fields
Mean trouble too. Each year the mastich yields
Fresh produce thrice, three times its berries grow.
Watch them and learn the times to reap and sow

(The early and the middle and the late). 1090
Plump bumper-crops of berries indicate
Abundant harvests; scanty berries mean
Lean harvests; pickings somewhere in between
Mean average yield. The three-times flowering squill
Also can show how fully crops will fill 1095 [1060]
Your reapers' carts. In fact, the squill displays
The progress of your crops in the same ways
As does the mastich.

Animal Signs of Foul Weather

When in Fall a swarm
Of wasps assembles, the first winter storm
Will form as swiftly as the wasps' whirlwind— 1100
Even before the Pleiades descend.
When the sow, ewe, and she-goat look for mates
Twice in a year, their strange lust indicates
Unseasonably early storms as well.
But ill-housed paupers cheer who can foretell 1105
A mild winter when their females mate
Once only in a year and do it late.

Cranes flocking past at the right time delight
Farmers who have discerned the time is right;
But cranes with deviant timing only cheer 1110 [1076]
Those plowmen who have no sense of the year.
Cranes always bring on winter—prematurely
When noisy flocks of them arrive too early.
Conversely, when the cranes come late in stray
Detachments, farmers bank on a delay 1115
In winter.
If the sheep and cattle root
In topsoil after you have picked the fruit
In fall or turn to face winds from the north,
The setting Seven Sisters will send forth
A wicked winter. If the cows and sheep 1120
Dig in the dirt too often and too deep,
Winter will drag on, stunting trees and crops.
Remember, though: whenever dense snow drops
Heavily on the fields and hides the stubble

And the stalks all are indistinguishable, 1125 [1089]
The prudent farmer looks for a good year.
May all the planets and the stars be clear
And only a few comets circle earth!
(Too many comets are a sign of dearth.)

The landlocked farmer shakes his hoe and spade 1130
When gorgeous flocks of island birds parade
Above his grain-fields at the end of spring—
He fears his ears will soon be withering
From drought. The grazier welcomes these same birds
In smaller echelons because his herds 1135
Promise to yield a wealth of milk that year.
Though every man must choose his own career
And hope that he can get ahead, we all
Cling to whatever omens rise or fall
And trust our lives to momentary signs. 1140 [1103]

The shepherd trusts in omens. He divines
A coming storm from herds that are too quick
To pasture or from nimble lambs that kick
With all four legs or hornéd rams that leap
Up on two legs and butt each other. Sheep 1145
Also foreshadow stormy weather when
They drag their sluggish hooves back to the pen
And stop to crop grass as their shepherds launch
Showers of stones.

Everyone on a ranch
Relies on cattle to foretell a squall. 1150
When cows lean on their right flanks in a stall
Or lick between their front-hooves, farmers know
To haul their harrows in. Oxen will low
Restlessly as they shamble home together,
And calves, too, somehow sensing rainy weather 1155 [1118]
Sooner or later will descend and soak
The range. When goats frisk near a prickly oak
Or sows run madly through the scattered hay,
A storm is on its way. When the wolves bay
And amble one by one down to the farm 1160
To make their lairs, when they fear no harm
From men, a storm will come within three days.

Here is a list of several other ways
To know when rains or storms will darken heaven,
Whether on the same day, the next, or even 1165
The third dawn:

deign to obey the mice—
They squeak more often when the weather's nice.
Our forebears never dared to disregard
When mice danced jigs or dogs dug up the yard
(For dogs dig when a rainstorm is at hand). 1170 [1132]
When hermit crabs come scuttling towards land
Expect a squall, and look for showers soon
When mice that scattered straw all afternoon
Cease from this practice and go home to nest.

Conclusion

Do not neglect these omens. It is best 1175
To take them as they come. If two agree,
Have doubled faith, but if you should find three,
Be certain. Watch these signs throughout the year,
Comparing whether the right days appear
When the right clusters rise and set. It's smart 1180
To watch the fourth day following the start
Of a new month and the fourth day from its end.
Beyond these boundaries lunar cycles blend
And months are old and new. The sky gives you
Scantier evidence on eight nights, too, 1185 [1152]
Because the yellow moon lies low. Assess
All of these signs as hours, days, weeks progress,
And never face the future with a random guess.

Appendix 1
Constellation Risings and Settings

In the northern hemisphere a number of circumpolar constellations such as Ursa Major and Minor never rise or set. The remaining constellations rise and set every day; however, we see only those risings and settings that occur between sunset and sunrise because daylight conceals the rest. Since the earth, in addition to rotating on its axis, is itself orbiting the sun, it daily changes its position relative to the stars, and the stars seem to move slowly westward, rising and setting roughly four minutes earlier each day. The sun therefore conceals a slightly different set of them in daylight each day. Because the stars return to the same positions relative to the earth on the same days every year, the rising or setting of a constellation just after sunset or just before sunrise serves as a marker indicating the progress of the year.

Technical terms for risings and settings of the constellations are as follows:

Morning rising—The rising of a constellation at the same time as or just before sunrise, also called "heliacal rising." On subsequent days the constellation rises earlier and therefore appears higher in the sky at sunrise.

Morning setting—The setting of a constellation at the same time as or just before sunrise. On subsequent days the constellation sets earlier, during the night.

Evening rising—The rising of a constellation at the same time as or just after sunset. On subsequent days the rising will take place earlier (in daylight) and the constellation will appear higher in the sky at sunset.

Evening setting—The setting of a constellation at the same time as or just after sunset, also called "heliacal setting." On subsequent days the constellation sets earlier (during daylight) and therefore is not visible after sunset in the night sky.

When conceptualizing these relationships, one should keep in mind that sunrise, though at the beginning of our day, is the end of night, and sunset, though at the end of our day, is the beginning of night. Furthermore, it is important to distinguish between "true" and "visible" risings and settings (see Van Der Waerden 1984, 107). A "true" morning rising, for example, would be one in which a star rises at the exact same time as the sun. This rising, however, would be invisible, since it would be hidden in the rays of the sun. A "visible" morning rising would occur a number of days later, when the star would be seen to rise just before the sun. "Visible" risings and settings are of use to the farmers and sailors whom Aratus repeatedly mentions, and he always refers to the times when the risings and settings can be seen.

As the Pleiades are especially significant in this context, let us take them as an example. After a period of invisibility that begins with their evening setting on April 1 and lasts for roughly forty days (Hesiod *Works and Days* 385–86), the Pleiades appear again in the night sky with their morning rising on May 14 and are then visible for roughly 320 days until the next April 1. After their appearance at the end of night (morning rising), they rise four minutes earlier each night until they eventually rise at the beginning of night on October 3 (evening rising). Their risings subsequently take place during daylight hours and therefore are not visible. Just over a month later, on November 13, the Pleiades, which had been setting invisibly during the daylight hours, set for the first time just before sunrise (morning setting). They then set four minutes earlier each night until they eventually set just after sunset on April 1 (evening setting). They subsequently set before sunset (while the sun is still shining), their roughly forty-day period of invisibility recurs, and the annual cycle begins again with their morning rising on May 14.

The morning setting of the Pleiades on November 13 was held to mark the beginning of winter. Aratus therefore can use this setting as a fixed sign by which to gauge an unseasonably early winter:

> When in Fall a swarm
> Of wasps assembles, the first winter storm
> Will form as swiftly as the wasps' whirlwind—
> Even before the Pleiades descend. (1098–1101)

Hesiod cites this same setting as a sign that the Mediterranean is no longer safe for seafaring:

> But if a desire for uncomfortable sea-faring
> Seizes you, truly gales of all kinds rage
> When the Pleiades drop into the misty sea
> To escape Orion's rude strength.
> (*Works and Days* 618–21)

Below is a list of the risings and settings that Aratus cites as seasonal signs. The dates are based on Geminos' *Calendar*, which compiles the *parapēgmata* of Callipos, Democritos, Dositheos, Euctemon, Meton, and Eudoxus (Aujac 1975, 98–108). Van Der Waerden's reconstruction of Euctemon's *parapēgma* was also helpful in compiling this list (1984, 105–6).

Rising or setting	Modern date	Lines in Aratus
Evening setting of the Pleiades	April 1	268–72
Morning rising of the Pleiades	May 14	268–72
Morning rising of Sirius	July 23	340–48
Morning rising of Vintager (Vindemiatrix)	Sept. 4	135–38
Morning rising of Arcturus	Sept. 14	796–99
Evening rising of the Goat (Capella)	Sept. 14	155–67
Evening rising of the Kids	Sept. 27	155–67
Evening rising of the Pleiades	Oct. 3	268–72
Evening setting of Arcturus (in the Plowman)	Oct. 29	602–11
Morning setting of the Pleiades	Nov. 13	268–72, 1099–1102
Morning rising of Scorpio	Nov. 24	311–13
Morning rising of the Eagle	Dec. 10	323–26, 547
Morning setting of the Goat (Capella)	Dec. 17	717–21
Morning setting of the Kids	Dec. 30	717–21

Appendix 2
Bayer Designations

After conducting thorough observations in both hemispheres, the German astronomer Johann Bayer published an innovative star atlas in 1603, the *Uranometria* (named after Urania, the Greek muse of astronomy). In this atlas he charted 1,564 stars, designating each with a Greek letter followed by the genitive form of its constellation's Latin name. Thus, Spica, in the constellation Virgo, was designated α Virginis (alpha of Virgo). As there are only twenty-four letters in the Greek alphabet, Bayer resorted to lowercase Latin letters for those constellations containing twenty-five or more stars. More often than not, he assigned α (alpha) to the brightest star in a given constellation and proceeded through the alphabet in order of brightness. In many instances, however, he assigned letters according to their location in the eastern, western, northern, or southern part of a constellation. Thus, for thirty of the eighty-eight constellations, α is not the brightest star.

I have keyed Aratus' relative directions to the Bayer system in the end notes. For convenience I list the Latin names, the Latin genitives and the English names of the constellations below. Also, because I have used Greek letters rather than their spelled-out form in English, I have provided the Greek alphabet and the transliterated names below.

Latin name	Latin genitive	Name
Andromeda	Andromedae	Andromeda
Aquarius	Aquarii	The Water-Bearer
Aquila	Aquilae	The Eagle
Ara	Arae	The Altar
Aries	Arietis	The Ram
Auriga	Aurigae	The Charioteer
Boötes	Boötis	The Plowman
Cancer	Cancri	The Crab
Canis Major	Canis Majoris	The Dog
Capricornus	Capricorni	The Goat
Carina	Carinae	The Ship's Keel (part of Argo)
Cassiopeia	Cassiopeiae	Cassiopeia
Centaurus	Centauri	The Centaur
Cepheus	Cephei	Cepheus
Cetus	Ceti	The Sea-Beast or Monster
Corona Borealis	Coronae Borealis	The Northern Crown
Corvus	Corvi	The Crow

Latin name	Latin genitive	Name
Crater	Crateris	The Mixing-Bowl
Cygnus	Cygni	The Swan
Draco	Draconis	The Dragon
Eridanus	Eridani	The River Eridanus
Gemini	Geminorum	The Twins
Hercules	Herculis	Hercules (On-His-Knees in Aratus)
Hydra	Hydrae	The Hydra
Leo	Leonis	The Lion
Lepus	Leporis	The Hare
Libra	Librae	The Scales (The Claws in Aratus)
Lupus	Lupi	The Wolf (The Beast in Aratus)
Lyra	Lyrae	The Lyre
Ophiuchus	Ophiuchi	The Serpent-Holder
Orion	Orionis	Orion
Pegasus	Pegasi	The Winged Horse
Perseus	Persei	Perseus
Pisces	Piscium	The Fishes
Piscis Austrinus	Piscis Austrini	The Southern Fish
Puppis	Puppis	The Ship's Stern (part of Argo)
Pyxis	Pyxidis	The Ship's Compass (part of Argo)
Sagitta	Sagittae	The Arrow
Sagittarius	Sagittarii	The Archer
Scorpius	Scorpii	The Scorpion
Serpens Caput	Serpentis Capitis	The Serpent's Head (part of the Serpent)
Serpens Cauda	Serpentis Caudae	The Serpent's Tail (part of the Serpent)
Taurus	Tauri	The Bull
Triangulum	Trianguli	The Triangle
Ursa Major	Ursae Majoris	The Greater Bear
Ursa Minor	Ursae Minoris	The Lesser Bear
Vela	Velorum	The Ship's Sails (part of Argo)
Virgo	Virginis	The Maiden

The Greek Alphabet

α = alpha	ι = iota	ρ = rho
β = beta	κ = kappa	σ = sigma
γ = gamma	λ = lambda	τ = tau
δ = delta	μ = mu	υ = upsilon
ϵ = epsilon	ν = nu	φ = phi
ζ = zeta	ξ = xi	χ = chi
η = eta	o = omicron	ψ = psi
θ = theta	π = pi	ω = omega

Notes

Invocation: 1–18

1–18 The invocation exhibits many aspects of a formal hymn: the god with whom the poet proposes to begin (1); a list of his attributes (2–13), and the invocation (14–18). In contrast to traditional literary hymns, reference is made to the god's arrangement of the cosmos rather than his mythic exploits.

1–3 In the Greek original (see Kidd 1997, line 2) Aratus puns on his own name with the adjective *arreton* ("unspoken") (Bing 1990, 291–95; Kidd 1981, 355; Levitan 1979, 68 n. 18). This pun serves as a sort of signature, deferentially placed after Zeus. It will not come over into English, but I have translated its meaning as "may our hymns / Never omit him." Zeus is here the Stoic *pneuma* ("breath") which pervades the cosmos and a providence which has rationally arranged the universe in such a way that men can make sense and use of it.

3–6 Aratus follows the associative logic of genealogical poems (e.g., Hesiod's *Theogony*): since mortals exist because of Zeus, he can be described as their father. Aratus takes Zeus' paternity of mortals a step further. In addition to being their originator, Zeus displays paternal affection and "kindness" towards them. St. Paul famously quotes part of line 3 in his speech to the Athenians (Acts 17.28): "We are his offspring, too." The image of a beneficent Zeus waking mortals for work in the morning opposes the less pleasant image in Hesiod's *Work and Days* 20.

6–8 Aratus distinguishes between rangeland for livestock (cows) and farmland (hoes). Agriculture here serves as a transition to the poem's astronomical theme: Zeus so arranges the cosmos that celestial bodies signal important times in the farmer's calendar.

9–12 These lines present the first of two origins for the constellations. Here, Zeus has grouped the stars in identifiable clusters to serve as signs for men. This account is hinted at again in reference to the Pleiades (269–72). In the alternate account (384–94) an anonymous human serves as "first inventor" of the constellations by dividing the stars into clusters.

14–18 These lines contain the formal invocation. The Greek word *chaire* ("hail") is a salutation. It can also be a valediction, however, and appears at the end of many of the *Homeric Hymns*. Scholars have disputed to what group "the elder race" refers: the Titans (the generation of gods before the Olympians), the elder brothers of Zeus (Poseidon and Hades), earlier astronomers, or the age of heroes. The phrase most likely refers, however, simply to the immortals in general, in contrast to the race of mortals. The Muses are *meilichiai* (simultaneously "honeyed" and "agreeable") like Zeus Meilichios, who protects worshippers who have invoked him with propitiatory offerings. Aratus introduces himself with "I"; elsewhere in the poem the first-person pronoun reemerges in emotional passages expressing wonder or dread (see note on 149–55 below).

The Axle, the Poles, and the Northern Sky: 19–332

19–25 Aratus' conception of the cosmos is, of course, geocentric. A central axle runs through the earth, which remains motionless. This axle extends to the top and bottom of the surrounding astral sphere, and the sphere rotates as the axle spins. The current northern polestar is Polaris (α Ursae Minoris). It is very close to the northern celestial pole.

25–45 Passing from the northern polestar to the circumpolar constellations, Aratus begins with Ursa Major (Greater Bear) and Ursa Minor (Lesser Bear). In prehistory seven bright stars in Ursa Major were identified as the Wain, or the Wagon (*hamaxa*). *Hamaxa*, consisting of *hama +axwn*, contains a pun on *axwn*, which Aratus uses for the "axle" of the sky (Kidd 1997, n. 27). This constellation had been renamed the Bear, however, some time before the eighth century BCE, when Homer mentions the Wagon as a second name for the Bear (H. *Il.* 18.487 = *Od.* 5.273). Since these original seven stars suggested only the body and tail of a bear, others were added. The less conspicuous and smaller Ursa Minor also originally consisted of seven stars and took its name from Ursa Major.

27–29 As is his custom throughout the *Phaenomena*, Aratus uses key stars to point out body-parts. The heads of the Bears are marked by α ο Ursae Majoris and β Ursae Minoris, respectively. Their tails by δ ε ζ η Ursae Majoris and ε δ α Ursae Minoris. The Bears' heads face in opposite directions, and each always points to the hind-quarters of the other.

30–35 Aratus here combines two myths: (1) that of Callisto, an Arcadian maiden and follower of Artemis, who, once she is ravished by Zeus and expelled from Artemis' company, eventually turns into the bear which becomes the constellation Ursa Major; and (2) that of the Goat Amaltheia who is said to have nursed the infant Zeus. The collocation of the phrase "if the tale's true" and Mounts Ida and Dikte in Crete evokes Epimenides' famous dictum: "All Cretans are liars." Aratus may be using this collocation to suggest that readers should not take his new myth seriously. He mentions the traditional version of Zeus' nurturing only 130 lines later (162–66).

At 8,057 feet, Mount Ida (now Psiloritis) is the highest mountain on Crete, standing in the center of the island. Mount Dikte, another prominent peak, lies in eastern Crete. Scholars have been unable to reconcile the distance between these peaks with Aratus' account.

34–35 "The elder men of Crete" are the Curetes, the masculine equivalent of nymphs; they are devotees of the mountain goddess Rhea and protect flocks and fields. Their lively and noisy dance, regarded as promoting fertility, served to disguise the infant Zeus' cries from his infanticidal father, Kronos.

36–45 Cynosure ("the Dog's Tail") is probably an earlier Greek name for Ursa Minor. Ursa Major takes the name Helíke (related to "helix") from her rotation around the pole. Odysseus navigates his raft by sighting Ursa Major at *Odyssey* 5.276–77: "The divine goddess Kalypso recommended that he sail with [Ursa Major] on his left-hand side." From 2000 BCE Ursa Major moved further from the pole and Ursa Minor nearer. According to Aratus' account, the Phoenicians, a seafaring and trading people, relied on her head star (β Ursae Minoris), which was near the pole in the first millennium BCE.

45–59 It is not as easy to disentangle the Dragon (Draco) from the Bears as Aratus states. The five prominent stars on his head are easiest to identify. The two stars that mark his temples are γ and ξ Draconis; β and ν mark his eyes, and μ his chin. His coils wrap around Ursa Minor, and the tip of his tail reaches to the head of the Great Bear. Prior to Aratus, Draco was simply called the Serpent. Aratus here disambiguates this constellation from the Serpent (Serpens) in the hands of the Serpent-Holder (Ophiuchus) (72–84). Ancient sources identify this long, coiling constellation with the snake of Ares slayed by Cadmus, the python slayed by Apollo, and the serpent that guards the golden apples of the Hesperides, slayed by Heracles.

60–61 There is a point in the north where some star clusters (the Dragon's head among them) reach their lowest point just above the horizon before beginning their eastern ascent again. These circumpolar constellations, such as the two Bears and the Dragon, never drop out of view. Most constellations, however, set at points on the western horizon and rise at points on the eastern horizon. The exact locations of these settings and risings vary according to a viewer's latitude.

61–69 "Silhouette" here translates *eidwlon* in Greek, meaning in this context either "form" or "ghost." Though no first-magnitude stars define this constellation, it does contain two bright globular clusters: M13 (the brightest in the northern hemisphere) and M92. Greek astronomers simply refer to this constellation as "On-His-Knees," or *Engonasin*. Later sources identify it as various heroes: Prometheus, Theseus, Tantalus, Thamyras, Ixion, and Heracles. Its modern name is the Latin form of Heracles: Hercules. Though most ancient astronomers do not speculate about the nature of this figure's labor, one late Scholiast (Q), 69, states that the figure is Heracles straining to hold up the weight of the sky. In Renaissance star maps the figure's right hand is depicted with a club to confirm the identification with Heracles/Hercules. The figure's right arm and hand consist of γ ω Herculis, and his left of μ ξ o; β marks his right shoulder, λ his left, and α his head.

69–72 The Crown lies just west of On-His-Knees and is best identified by the parabola of stars θ β α γ δ ε ι Coronae Borealis. It is now called Corona Borealis to distinguish it from Corona Australis in the southern sky. According to Aratus, Dionysos sets the crown in the sky as a *sēma* (root of the English word *semiotics*) for Ariadne, a pun conveying that the crown is both a sign in the sky and a memorial (funerary) marker for Ariadne. Aratus here handles the myth summarily on the assumption that it is well-known to his audience. After Theseus abandons Ariadne on the island of Naxos, Dionysos arrives in state, with his retinue, on a chariot drawn by panthers, and takes her as his wife. As she is mortal, she eventually dies.

72–84 "Serpent-Holder" is a literal translation of the Greek *Ophiouchos*. The constellation is known today by the Latin transliteration of the Greek name, Ophiuchus. Aratus advises his addressee to locate the head star of Ophiuchus (α Ophiuchi) from the head of On-His-Knees (α Herculis). β γ Ophiuchi designate his right shoulder, and κ Ophiuchi his left. The "mid-month moon" is the full moon, which, according to Aratus, does not shine brightly enough to obscure the Serpent-Holder's prominent shoulder stars. The stars which mark his hands are shared with the Serpent (Serpens): ν Ophiuchi for the right hand, and δ ε for the left. His body, though indistinct, divides the Serpent into two clusters that are counted as one constellation, Serpens Caput (Snake Head) and Serpens Cauda (Snake Tail). His right-foot star is θ Ophiuchi, and his left-foot stars, closer to the Scorpion (Scorpio), are ψ ω ρ. As the Scorpion belongs to the southern sky and is a prominent and familiar constellation, Aratus introduces it only in passing here. In the modern conception δ π Scorpii mark its eyes, ν β its left claw, and ρ its right claw. Its tail begins with ε Scorpii and ends with the stinger (ι κ λ υ). The Scorpion figures in a later mythic narrative involving Artemis and Orion (668–83).

84–86 With the formation of the twelve-sign Zodiac, the Claws became a constellation disassociated from the Scorpion. The modern name for this constellation—the "Scales of Balance" (*Zugon* in Greek, *Libra* in Latin)—develops later, possibly due to the proximity of Virgo (associated with Themis, goddess of justice). One scale in the balance (one claw for Aratus) consists of θ γ Librae, and the other of υ τ. The triangle β α σ binds the two together. The Claws (Libra) lie on the ecliptic (the sun's path), and the autumnal equinox (Sept. 22 or 23) occurs when the sun is in Libra.

87–92 The constellation Arctophylax has an earlier name, Boötes (the Plowman), which dates from the time when the Greater Bear was known as the Wagon (see note on 25–45), and this name is now its modern name as well. When the Wagon became known as the Bear (Arktos), the bright star nearby became known as Arcturus or Arktophylax (the Guardian of the Bear); by Aratus' day, the whole constellation had taken on the name Arctophylax. In Greek mythology, Arcturus is a star created by Zeus to protect the nearby constellations Arcas and Callisto (Ursa Major and Ursa Minor). Arcturus is the third-brightest star in the heavens, after Sirius and Canopus. Astronomy instructors advise their students to follow the handle of the Big Dipper (the seven brightest stars of Ursa Major) and then "arc to Arcturus and speed on to Spica" in order to get their bearings. For Spica, see note on 93–97 below.

β Boötis marks the Plowman's head, and γ δ his left and right shoulders, respectively. His left arm (λ) stretches back to his left hand (κ θ), which is holding a whip. α Boötis (Arcturus) marks his left knee and ξ his right.

93–97 The Maiden is the constellation Virgo. Spica (α Virginis), a star of the first magnitude, designates the ear of grain (for which *spica* is the Latin word) which she is holding in her left hand. β Virginis marks her head, and η her neck. The name Maiden (*Parthenos* in Greek) most likely derives from this cluster's association with the cult of the grain-bearing goddesses, Demeter and Kore (cf. Erren 1967, 38). Aratus' "multiple-choice" paternity for the Maiden ("Astraeus . . . or some god else," 95–96) is in keeping with the Hellenistic interest in mythical origins, especially when there are competing traditions (Kidd 1997, n. 98). Astraeus also appears in Hesiod's *Theogony* 378–82: "By Astraeus Dawn gave birth to the winds and the gleaming stars which heaven crowns." It is not clear how this god who "first sired the stars" (mentioned only here in Aratus) relates to the Zeus who originally arranged the stars in the sky (9–12). We see a similar mythological overlap elsewhere in the poem (for example, 31–35 and 162–63), and these overlapping accounts hint at the poet's awareness that not too much credence should be given to them.

97–135 Aratus cites a tradition ("some maintain . . . ," 97) according to which this constellation is the goddess Justice, or Dike. The subsequent narrative evokes Hesiod's Five Ages of Man (*Works and Days*, 109–201) and Dike (*Works and Days*, 220–62). In both Hesiod's and Aratus' Golden Age, gods and mortals freely interact. In Hesiod's account, humankind violates the earth and the sea with agriculture and sailing by forcing itself, with plow and keel, into regions that were not meant for it—man was meant to walk on top of the earth. Agriculture and sailing are thus regarded as signs of devolution. Though humankind does not sail in Aratus' Golden Age (106–7), Aratus does include domesticated animals (cows and plows), and mortals therefore cultivate the earth. According to Stoic doctrine, all human occupations were as old as humankind itself (Kidd 1997, n. 112). Aratus, however, "placed agricultural labor in the Golden Age not just in deference to Stoic doctrine, but because agriculture is itself a manifestation of divine ordering and justice" (Fantuzzi and Hunter 2004, 240). There are markets for domestic goods (101) but, as yet, no goods imported through oversea trade. At least partly due to each community's self-sufficiency, there is no war in the Golden Age.

Before the Silver Age the other gods withdraw to heaven, and Justice retreats from the city to foothills and mountains. In the Bronze Age humankind gives up its pacifism and vegetarian diet and becomes violent and carnivorous (128–31). By turning from the exclusively agricultural pursuits of the earlier ages, the men of the Bronze Age cease operating in accordance with (Stoic) divine order, and "much more is destroyed than just animal life" (Fantuzzi and Hunter 2004, 240). Justice then withdraws to the heavens in disgust, and Aratus concludes his account of the Ages of Man with her catasterism, omitting Hesiod's Ages of Heroes and of Iron. The Age of Heroes was regarded as "historical" by Hesiod (and the ancient Greeks in general)—that is, as the period during which the events of the *Iliad*, for example, took place. Hesiod's Iron Age takes place during his narrative present: it is "right now." Aratus most likely omitted these two Ages because "a message of progressive decay would in fact hardly suit the rest of the poem" (Fantuzzi and Hunter 1994, 240). Though the men of Hesiod's various Ages die in violent or mysterious ways, Aratus also does not explain how the men of his Ages die off. Each of his three Ages are much more recognizable as similar to "our time" than Hesiod's, and divinities have no effect on them; the gods do nothing more than withdraw from the world.

Kidd has identified a triadic structure in this passage—Strophe, Antistrophe and Epode: two fourteen-line sections (97–111, 112–26) pertain, respectively, to the Golden and Silver Ages, and the third, a shorter section (127–32) pertains to the Bronze Age (Kidd 1997, n. 101).

115 The "twilight" setting contributes to the mood of decadence. The only sign of devolution in the Silver Age, however, is Justice's withdrawal to the mountains. It is ironic, then, that with her complete retreat from the earth and eventual catasterism she again becomes visible to all humankind (133–35). Kidd (1997, n. 135) traces "a kind of progression in the visibility of Dike" from day to dusk to night that is linked to her progression from cities, to mountains, and to the sky.

124 Hesiod (*Works and Days*, 181) prophesies that the Iron Age will end when men "are born with grey hair around their temples," that is, when anxiety increases to such an extent that even a fetus will feel it in the womb. Justice here predicts an increase in anxiety during the Bronze Age.

132 Justice leaves the earth in disgust, just as, we are told in Hesiod's *Works and Days* 197–201, Aidos (Shame) and Nemesis (Righteous Indignation) will leave the world in the sixth and final Age: "And then Aidos and Nemesis, their sweet forms wrapped in white robes, will depart from the wide-pathed earth and leave humankind to join the company of the immortal gods. And bitter sorrows will remain for mortal men, and there will be no way to ward off evil." Whereas Hesiod leaves the reader with an apocalyptic vision, Aratus goes no further than meat-eating and weapon-making.

133–35 Hutchinson (1988, 223) points out that "the final detail of Virgo's position recalls the original introduction of the constellation [92–93]." Thus, the pedantic descriptions of a clearly observable constellation encapsulate, by ring composition, a myth that details the Maiden's retreat from the sight of men. According to Hutchinson, this abrupt return to astronomy "mingles the prosaic with the subversive" (224).

135–38 Editors are generally agreed that the line in the manuscripts that refers to the wings of the Maiden (line 138 in Kidd) is an interpolation. However, since later images of this constellation include wings and Kidd retains the line in brackets, I have opted to translate it. γ and δ Virginis mark the Maiden's left and right sides, respectively, and the bright star along her right wing is Vintager, or ϵ Virginis, named for its morning rising on September 4, the time for grape gathering (Geminos' *Calendar*, in Aujac 1975). Aratus compares the magnitude and brightness of ϵ Virginis to those of another unaffiliated star, α Canum Venaticorum. As third-magnitude stars, they are both comparatively bright. Though naked-eye astronomers usually do not distinguish between a star's magnitude and its brightness, Aratus does refer to the amount of space a star occupies in heaven and its brightness as distinct qualities (Kidd 1997, nn. 139, 140).

139–46 Aratus uses the Great Bear's paws (β and γ Ursae Majoris) as a guide to a group of stars that do not belong to a constellation. He points out similar clusters of "amorphous" stars at 380–401 ("The Anonymous Stars") and at 407, as he has done with the unaffiliated star ϵ Virginis above. It is striking that Aratus applies the epithet "marvelous" *(deine)* both to the Bear and the anonymous, amorphous stars beside her. As Hutchinson (1988, 218) observes, "The disparity between physical glory and mythological obscurity yields a strange kind of half-pathos for the inanimate in their inanimation." Eudoxus, Aratus' source here, does not stress the fact that these stars are unaffiliated (Eudoxus fr. 28).

θ Ursae Majoris marks the Great Bear's forelegs and κ ι her forepaws; her hind legs are separated left (ν ξ) and right (λ μ); and ψ marks her hind knee. The unaffiliated stars before her forelegs were subsequently gathered into the small constellation the Lynx, and those before her hind legs were gathered into another small constellation, Leo Minor.

147 The two stars α β Geminorum mark the heads of the Twins. In antiquity the twins were identified as Apollo and Heracles, Amphion and Zethus, and Triptolemos and Iasion (attendants and favorites of Demeter) but most often as Castor and Pollux, who become the modern Gemini and give their names to the head stars, α and β, respectively. η μ ν Geminorum mark the feet of Castor, and γ ξ those of Pollux.

148 Aratus here summarily introduces the Crab (Cancer). This constellation will figure in later accounts of the risings and settings, since the summer solstice reaches its most northerly point in this constellation (the Tropic of Cancer) (521–22). The faintest of the Zodiac constellations, the Crab is best identified by a slightly bent line of stars running α δ γ ι Cancri. α and ι represent its left and right claws, respectively. δ and γ mark the Crab's left and right eyes, respectively, and lie on either side of the Tropic of Cancer (515–17). The Crab contains a visible open star cluster known as the Manger (now M44), and δ and γ Cancri also take the name "the Asses" from their proximity to it. Aratus will later explain how variations in the appearance of the Manger and the Asses predict changes in the weather (937–53).

149–55 Like the Crab, the Lion (Leo) lies on both the Tropic of Cancer and the ecliptic. The hottest weather occurs when the sun is in Leo, after the summer solstice (on or near June 21) and after the grain has been harvested. The Etesian (or "Yearly") winds are northerlies which occur in the eastern Mediterranean. Aratus interjects a prayer in the first person that expresses his fear of sailing during this season. Though he does not specify the season, Aratus' model Hesiod also expresses an abject fear of sailing (*Works and Days* 646–62). As Aratus uses Leo's body parts to define the northern tropic later in the poem (513–15), it is prudent to provide its key stars here: μ ε λ κ Leonis mark the Lion's head, and β his tail. α θ ι mark his breast, belly, and genitals. Ancient sources often refer to α Leonis, or Regulus, as "the heart of the Lion."

156–67 Aratus uses the Twins and the Great Bear to guide the reader to the Charioteer (Auriga) as well as the Goat (the star Capella) and the Kids (three stars near Capella), which he contains. Several ways of combining the stars of this constellation into an image are current. In one, the constellation represents only the Charioteer's pointed hat (δ Aurigae) and head (α ε ζ η Aurigae serving for a nose, and β θ ι filling out the rest of the head). Aratus makes clear later in the poem, however, that he regards Auriga as full-bodied figure. In the traditional version, δ Aurigae marks his head, α β his shoulders, η θ his hands, and ι Aurigae and β Tauri his feet. The Charioteer was identified with several mythological figures: Erichthonios, the autochthonous king of Athens, who is said to have invented the chariot; Orsilochos, an Argive who invented the four-horse chariot; Myrtilos, the charioteer of King Oenomaus in his race with Pelops; and Phaethon, who disastrously takes the reins of his father the sun god Helios' chariot.

The Goat is α Aurigae (Capella), and the Kids ε ζ η. The Goat and the Kids were important because they were believed to forecast stormy weather with their risings in mid- and late September. (On constellation risings and settings see Appendix 1.) Ancient sources propose a number of possible origins for "Olenian." The most likely is the Greek noun *ōlenē* (meaning shoulder), since α Aurigae marks both the Goat's and the Charioteer's shoulders. In myth this goat is Amaltheia, who nurses the infant

Zeus on Crete (162–66). Aratus has already informed us, however, that two she-bears suckle Zeus instead—a version unattested elsewhere (see lines 31–35 and note on 30–35 above).

167–79 A distinctive V of stars called the Hyades designates the head of the Bull (Taurus): γ Tauri is the point from which ε β Tauri extend for his left horn and α ζ (Aldebaran) for his right. γ Tauri marks his right horn-tip, and β Tauri serves as both the right foot of Auriga and the left horn-tip of Taurus. Since, in his description of Achilles' Shield, Homer refers to this entire constellation as the Hyades (*Iliad* 18.486), it is likely that the name "the Bull" (Taurus) came into circulation after the eighth century BCE. In antiquity the Bull was identified with several mythological bulls: Zeus in disguise seducing Europa, the Cretan Bull (which is later the Bull of Marathon), and even Io (who was transformed not into a bull but a heifer).

In myth the Hyades ("the Rainy Ones") are the nymph daughters of the Titan Atlas and sisters of the Pleiades (260–72). When a lion kills their brother, Hyas, they weep until they become a misty cluster in the heavens. Hyas then becomes the constellation Water-Bearer (Aquarius). The Hesiodic *Astronomy* (fr. 291) gives the names for five Hyades: Phaesyle, Coronis, Cleeia, Phaeote, and Eudora.

179–204 Aratus assumes a tragic tone for his description of the Cepheus group—Cepheus, Cassiopeia, and their daughter, Andromeda. First performed in mid- and late fifth century BCE, the *Andromeda* plays of Sophocles and Euripides were greatly popular in Athens and, eventually, throughout the Greek-speaking world. These plays, unfortunately, have not come down to us. It is likely that the names for these three constellations were coined during this period or shortly thereafter. Aratus refers to Cepheus as "Iasides," or the son of Iasius, who was himself descended from Io and the Egyptian king Belus. This patronymic, as it calls to mind Io, who bore a son, Epaphos, to Zeus, serves to keep Zeus' influence in the background of the poem.

Cepheus was the king of the Phoenician nation of Aethiopia. Its capital was the seaport Joppa (modern Jaffa, the port of Tel Aviv). When his queen, Cassiopeia, boasts that she is more beautiful than the Nereids, Poseidon sends a Sea-Beast to destroy Joppa. Cepheus offers Andromeda as sacrifice. Just as the beast is about to kill Andromeda, the hero Perseus arrives, slays the beast, and takes Andromeda as his wife. Perses, the son of Perseus and Andromeda, goes on to found the Persian race.

Cepheus holds his feet (γ κ Cephei) towards the North Pole. Other third-magnitude stars designate his belt (β Cephei), shoulders (α ι), and head (ζ). Aratus specifies that the stars marking the tip of Cynosure's tail (Polaris) and Cepeheus' left and right feet form an equilateral triangle (compare the lines on the constellation Triangulum, 237–43).

The five bright stars β α γ δ ε Cassiopeiae form the W-shape that constitutes the most conspicuous part of Cassiopeia. The constellation is that of a female figure on a chair, though Aratus does not mention the chair until he introduces Perseus (257). Interpretations of the lines containing the elaborate simile vary widely, and scholars dispute what the constellation is being compared to. Many argue that it is compared to bosses on an opened double-leaved door. I have adopted this reading in the translation. On the basis of evidence from Homer's *Odyssey*, Kidd, in contrast, argues that the constellation is compared to a temple key (larger and more angled than modern keys). β and α Cassiopeiae designate her right and left shoulders, respectively.

The constellation Andromeda is conceived as a figure with arms outstretched and enchained at the moment of her sacrifice to the Sea-Beast. α Andromedae marks her head, δ and σ her left and right shoulders, respectively. γ marks her left foot, and φ her right. β μ ν Andromedae designate her sash. The same stars mark her hands and bonds simultaneously, ι κ λ for the right and ζ η for the left.

205–27 Subsequently in antiquity and throughout the modern era the Stallion was identified with Pegasus, the winged steed of the hero Bellerophon. Though Aratus does not assign wings to this constellation, he does mention Pegasus broaching the Hippocrene. Andromeda's head-star (α Andromedae) also marks the belly of the Stallion (Pegasus). This star and three others form the quadrilateral outline by which the constellation is most easily identified: α Pegasi marks its shoulders, β the forequarters, and γ the haunches. ε Pegasi, then, marks its nostrils, and θ an eye. Helicon is one of the traditional haunts of the Muses and the location of Hesiod's poetic initiation (*Theogony* 1–115). "Hippocrene" means "horse-fountain" (*hippo* + *krēnē*). The ancient city of Thespiae was midway between Mount Helicon and Thebes to the east.

227–36 The Ram (Aries) lies on the celestial equator. Since it is much farther from the pole than the Lesser Bear (Cynosure), it revolves more rapidly. Andromeda's sash (β μ ν Andromedae) points to the three stars which best mark the constellation (α β γ Arietis). These stars designate his head and left horn. Orion's belt (ζ ε δ Orionis) does lie on the celestial equator, but Scorpio's pincers (ν β and π ρ Scorpii) are 20–25° south of it. See 534–47 and the note below.

237–43 Aratus takes great delight in poeticizing geometrical concepts. *Triangulum* is the Latin and modern name for the constellation Aratus calls *Deltoton* ("shaped like the Greek letter capital Delta"—i.e., Δ). β γ Trianguli mark the short line of the isosceles, and α the apex. In antiquity this constellation was also identified with the Nile Delta and the Island of Sicily.

243–53 The Fish (Pisces) consists of two separate fish joined by cords, and α Piscium marks the "Knot of Heaven" (or "Knot of Tails") where the cords come together. Andromeda's left shoulder (δ Andromedae) is near the northern Fish (τ υ φ ψ Piscium). As this Fish is higher above the celestial equator, it is more useful for Aratus' audience in the northern hemisphere. η Piscium lies midway between the northern Fish and the Knot. A distinctive circlet of fourth- and fifth-magnitude stars (ι θ γ κ λ TX Piscium) defines the southern Fish, which lies just above the celestial equator. Neither of these two fish is to be confused with the constellation Southern Fish at 397–401.

Ovid provides a summary account of these Fishes' association with Dione (Aphrodite) and Cupid in myth:

> . . . Pisces, ethereal steeds. They say you and your brother (for your stars shine together) carried two gods on your backs: once on a time, while running from terrible Typhon (back when Jupiter bore arms for the sake of the skies), Dione reached the Euphrates with tiny Cupid and sat on the margin of Palestine's stream. Tall poplar and reed covered the banks; willows, too, were giving hope that

> they might be concealed. While they hid, the wood roared with the wind. She grew pale with fear and thought that a hostile band was approaching. Holding her son to her breast, she cried: "Nymphs, come to our rescue; give help to two divinities." There was no delay; she leapt. Twin fish emerged beneath them. As you see, the present stars are named in memory of them. (*Fasti* 2.458–72)

Eratosthenes relates an alternate myth in which these fish were set in the heavens in commemoration of the fish that saved the Syrian goddess Derceto, who had flung herself into Lake Bambyce, near the Euphrates (*Catasterisms* 38).

253–59 After successfully obtaining the head of the Gorgon Medusa, Perseus, son of Zeus and Danaë, rescues Andromeda from the Sea-Beast in Joppa and takes her as his wife. τ Persei marks his head, and γ and θ his right and left shoulders, respectively. η marks the right hand gesturing toward Cassiopeia's chair, α his belt, ϵ ξ ζ his left leg, and δ μ his right. The phrase "his father Zeus' property" operates on several levels: Zeus is simultaneously Perseus' father, the Homeric "father of gods and men," and the Stoic origin of all life. The sky itself (his "property") is divine, inasmuch as it is pervaded by Zeus.

260–72 Though an unattached group of stars in Aratus, the Pleiades have belonged to the constellation Taurus since the time of Ptolemy the astronomer (83–168 CE). Aratus here gives ϵ Persei (Perseus' left knee) as a guide to them. Though some ancient authorities claim to have discerned all of the traditional seven stars or even more, Aratus contends that only six can be seen. (When the Fool asks Lear "why the seven stars are no mo' than seven," he smartly answers, "because they are not eight." *King Lear* 1.5.28–30.) In modern reckoning, however, there are nine named stars: the seven sisters that Aratus lists here out of a half-serious deference to poetic tradition and then two stars named for their father, Atlas, and mother, Pleione (see the section on Hyades at lines 171–74 and the note on 167–79). Their risings and settings at sunrise and sunset are especially important as agricultural and seasonal signs; for a full account, see Appendix 1.

272–79 The Lyre (Lyra) is a small constellation best identified by its brightest star, Vega (α Lyrae). On-His-Knees' left knee (θ Herculis) stands on one side of the constellation, and the Swan's head (β Cygni) on the other. κ β γ θ Lyrae define the shell of the Lyre, and α ϵ δ its strings. In *Homeric Hymn* IV the infant god Hermes, on the first day of his life, catches a tortoise, hollows out the shell, and with the addition of animal hide and antelope horn, makes the first lyre.

> As swift thought darts through a man's breast when thronging cares beset him, or as brightness flashes from his eyes, so glorious Hermes came up with both plan and act at once. He cut stalks of reed to the right length and made them secure by fastening their ends across the back and through the shell of the tortoise. He then skillfully stretched ox-hide all over it. He also added the horns, fitted a cross-piece over the two of them and then stretched seven strands of sheep-gut. When he had finished it, he held the lovely thing and tuned each string in turn with the key. Under his touch it sounded marvelous. While trying it out, the god sang sweet

> random phrases, just as young men taunt one another at festivals. (*Homeric Hymn* IV.43–50)

Later that day Hermes steals the god Apollo's cattle and eventually gives the lyre to Apollo as recompense for the theft.

279–86 The constellation described here, called "the Bird" in Aratus, becomes associated later in antiquity with the mythological Swan (Zeus in disguise) that rapes Leda; it is now known as Cygnus the Swan. The "middling vivid" stars β η γ α Cygni mark his head, neck, center, and tail, respectively. In Aratus' conception, the observer sees the underside of the bird. δ ι κ Cygni, then, constitute the right wing, which brushes Cepheus' hand, and ϵ ζ the left wing near the Stallion.

287–92 Aratus here follows two constellations of the Zodiac, the Water-Bearer (Aquarius) and Capricorn, into the southern celestial hemisphere. The Water-Bearer's waist consists of θ ι Aquarii, and his right and left shoulders of α and β, respectively; μ ϵ mark his left hand. γ ζ η π Aquarii, which mark both the Water-Bearer's right hand and the Water-Jug, are near θ Pegasi in the Stallion. Aratus describes the water poured from this jug as a separate constellation, the Water, in lines 401–15.

Capricorn consists primarily of the bent line of stars α β ρ ψ ω Capricornis, which runs from his head to his front hooves. Aratus never describes this constellation; in art, however, Capricorn is usually an amphibious half-goat, half-fish creature adapted to its proximity to the ocean. γ δ Capricornis, then, represent the constellation's termination in a fish tail. The "lowest track" (291) is the Tropic of Capricorn (23.5° S). The winter solstice occurs when the sun reaches this tropic, its furthest point south. For more on this tropic, see 523–34 and note.

293–308 Aratus again introduces himself into the poem with a first-person pronoun and addresses the reader with a "you" for the first time (rather than the "you" understood in his imperatives). Zeus makes his first appearance here as a cruel weather god. Briefly leaving behind his broad cosmic view and Stoic *ataraxia* ("tranquility"), Aratus here speaks like his model Hesiod as an individual whom storms at sea can terrify. During winter on the Mediterranean strong southerly winds are common; the Romans referred to the Mediterranean as the *clausum mare* ("the closed sea") during this time.

Though the identification of the bird referred to here is uncertain, it is generally accepted to be a shearwater, a seabird with species in the *Puffinus* and *Calonectris* genera. Lines 293–301 describe the weather during midwinter, when the sun is in Capricorn (see Appendix 1). At 302 Aratus broadens his description to the entire winter.

308–19 The Archer (Sagittarius) is most easily identified by his bow (μ λ ϵ η Sagittarii) and arrow (ζ δ γ). This arrow is not to be confused with the separate constellation, the Arrow (320–22). Aratus envisages Sagittarius as a centaur, as is the common conception: α β Sagittarii mark his right front fetlock and hoof, respectively; θ marks his right rear thigh; and ι his right rear hoof. In myth Sagittarius is identified with Crotus, the son of Pan, who consorted with the Muses on Mount Helicon and invented the hunting bow and the percussive rhythms which accompany music.

Aratus has described above the dangers of the sea when the sun is in Capricorn (293–308). He now backtracks to the beginning of the prior month to provide seasonal signs of stormy weather. When the morning rising of the Scorpion has occurred (Nov. 24) and the sun is in Sagittarius, winter is at hand, and storms at sea will make sea travel dangerous (Geminos' *Calendar* in Aujac 1975). Aratus advises that it is prudent to desist from sea travel at this time. At the morning rising of the Scorpion, Cynosure's head (β Ursae Minoris) is nearest to the North Pole, Orion sets in the west, Cepheus' right hand (η θ Cephei) descends, and the whole constellation Cepheus disappears down to the waist (β). Euctemon notes the rising of the Scorpion's sting on December 4 (Geminos' *Calendar* in Aujac 1975). For a more detailed account of constellation risings and settings, see Appendix 1.

320–22 Four stars define the Arrow (Sagitta): γ Sagittae marks the head, δ the shaft, and α β the fletched end. In myth this arrow is one of those shot by Apollo, Heracles or Eros (later Cupid).

323–26 Three stars in an echelon formation, suggesting a bird in flight, constitute the core of the Eagle (Aquila), γ η λ Aquilae. The Eagle is the bird of Zeus; Aratus mentions this association at 547. Aratus accurately asserts that the Eagle lies on the celestial equator. In fact, θ and η Aquilae, which are very near one another, lie on either side of it. The Eagle's morning rising occurs on December 10 during the stormy winter months.

326–29 The Dolphin (Delphinus) is a small constellation in the shape of a slightly irregular parallelogram (γ α δ ζ Delphini) with a tail (ϵ). In myth, a certain Delphinus is said to have persuaded the Nereid Amphitrite to accept the advances of the sea god Poseidon. In recompense, Poseidon sets the constellation that bears his name in the heavens. In a different legend, the poet Arion of Lesbos (seventh century BCE) escapes from a crew of sailors who had threatened to kill him and take his goods by jumping overboard and riding away on a dolphin that had been charmed by his music (Herodotus *Histories* I.23–24). This dolphin was then set in the heavens as a reward for its good deed.

329–32 In these transitional lines Aratus elegantly closes his survey of the northern constellations and opens his survey of the southern ones. He has, however, already introduced two southern Zodiac constellations, the Water-Bearer (Aquarius) and Capricorn, above (287–92).

The Southern Sky: 333–474

333–36 Aratus provides a summary account here of the large and prominent constellation Orion. λ Orionis marks his head, and α and γ mark his right and left shoulders, respectively. His right arm (μ) reaches above his head, and his right hand (ξ ν) holds a club (χ^2 χ^2). Three second-magnitude stars (ζ ϵ δ) make his belt especially conspicuous. Aratus provides his mythological background in lines 668–83. In Homer's *Odyssey* (11.572–75) Orion hunts wild beasts with a club even in the Underworld. His character later conforms to the constellations around him: he becomes a hunter with a sword (and sword belt) instead of a club, with the Dog (Canis Major) as his companion (337–50), and together they hunt the nearby Hare (Lepus) (350–54).

337–50 The Dog (Canis Major) takes its name from the alternate name for Sirius (α Canis Majoris), the "Dog Star," which is the brightest in the heavens. Eventually, stars near Sirius were gathered into a dog-shaped constellation, with the dog rearing on its forelegs, below and behind Orion: ζ Canis Majoris marks its hind legs; β its forelegs; ϵ σ δ o^2 π its middle; and η ω τ its tail. Its head consists of θ γ ι with Sirius (α Canis Majoris) at the tip of its muzzle. To provide the etymology of the name Sirius (*Seirios* in Greek), Aratus uses the verb *seiriaei* ("to scorch," "to sear"). By mere chance, this etymology comes over into English as a pun.

Sirius is especially important because its morning rising (July 23) occurs during the hottest time of the year (Geminos' *Calendar* in Aujac 1975). In 348–49 Aratus is more likely referring to the morning setting of Sirius (Dec. 6) as a storm sign in winter than to the evening setting (Apr. 26).

350–54 Placed under (south of) Orion's feet (β and κ Orionis), the Hare (Lepus) was imagined to be facing west, as if running from the Dog. ϵ Leporis marks its forelegs, α β its body, γ δ its hind legs, and μ its head. As the Hare is west of the Dog, it sets first.

355–64 In myth the *Argo* is the ship in which Jason and the Argonauts sail to retrieve the Golden Fleece from the eastern coast of the Black Sea. Athena assists the shipwright Argos in the construction of it, and the hull contains a divine plank from the oracular oak at Dodona, which gives the ship the power of speech. The Argo is the largest constellation in the southern sky and is divided into four components in modern star charts: Carina (Keel), Puppis (Stern), Pyxis (Compass), and Vela (sails). No stars mark the prow, which is often omitted in representations. γ α β Pyxidis define a mizzen mast instead of a compass in Aratus' conception. π ν τ Puppis define the tiller which dangles behind the Dog. Aratus uses a simile describing the stern-first beaching procedure of an ancient ship to account for the reverse movement of the Argo through the heavens.

364–370 The Monster, or Sea-Beast (Cetus), is the creature sent by Poseidon to kill Andromeda. She was offered as a sacrifice in retribution for her mother Cassiopeia's boast (see the note to 179–204, above). This large but comparatively amorphous constellation, lying beneath the Ram and Fish, consists of three groups: α γ ν ξ μ λ Ceti for the head, ζ θ η τ for the body, and ι as the tail. Aratus does not include β Ceti as the tip of the tail but assigns it to the Water (407–12). Aratus later uses the Knot of Tails (α Piscium) to point out the nape of the Monster's neck (behind δ Ceti) (373–77). Since it rises after Andromeda, the constellation conveniently "pursues" Andromeda in the heavens as the Sea-Beast does in myth. Its head stars lie above (north of) the river Eridanus, which Aratus introduces as the next constellation.

370–73 The Eridanus is a mythological river, now associated with the Rhine, the Rhone, or the Po. Phaethon, son of the god Apollo, is said to have lost control of his father the sun god's chariot and dried up the Eridanus by flying too close to it. Aratus waxes tragic in his description of this constellation and may be alluding to Euripides' partially preserved tragedy *Phaethon*. The star that marks the beginning of the river (λ Eridani) is very close to Orion's left foot (β Orionis); from there it trends west and then east again before ending at α Eridani. Eridanus "trickles at the feet of gods" (372) because the sky is conceived as the floor on which the gods walk. In a display of digressive variation technique, Aratus backtracks to explain that the Knot of Tails (α Piscium) lies near the top of the Monster's spine.

377–97 This brief digression exhibits a chiastic, or "mirror-image," rhetorical structure common in Hellenistic verse:

(a) 377–81: unnamed and ungrouped stars
(b) 382–83: named and grouped stars
(c) 384–92: the first inventor of the constellations
(b) 393–94: named and grouped stars
(a) 395–97: unnamed and ungrouped stars

Aratus has informed us in the invocation that Zeus set clustered stars in the heavens to serve as signs for humankind (9–12), but he makes no mention of the unnamed and ungrouped stars there. In this passage a first inventor of the constellations (himself unnamed) assigns names and figural shapes to the clustered stars. Some stars, however, remain unaffiliated with a constellation; they are of no use to humankind.

According to this "first inventor" theory a single individual came up with all the constellations at the same time. This theory is, of course, incorrect. Some constellations (and their names) evolved from Babylonian origins; others (e.g., the Hare) first appear in Aratus' source, Eudoxus. Aratus himself is aware of earlier names for some constellations (e.g., the Wagon as an alternative for the Great Bear) and most likely is using the first inventor simply as a way to account for the unnamed stars.

397–401 The Southern Fish (Piscis Austrinus) is southeast of Capricorn and southwest of the Water-Bearer. According to the Greek conception, the directions take their names from the winds which blow from them; thus, Notus (the wind from the south) blows upon the Southern Fish. Though Aratus does not point out specific stars in this constellation, his inclusion of the star Fomalhaut (also known as α Piscis Austrini) in the Water (410) indicates that his conception differs from the modern one. In the modern conception Formalhaut marks the fish's mouth and ι Piscis Austrini its tail.

401–15 The Water is not one of the modern constellations. It here takes its name from appearing as if it were poured forth from the right hand and pitcher of the Water-Bearer (γ ζ η π Aquarii). There are two bright stars in this otherwise dim constellation: α Piscis Australis, or Formalhaut (which Aratus does not include as part of the Southern Fish), lies due south of the Water-Bearer's right foot (δ Aquarii), and β Ceti (which Aratus does not include as the tip of the Monster's tail). Aratus includes in this constellation a group of stars further west, south of Sagittarius (413–15). This group later becomes part of the modern Southern Crown (γ α β δ Coronae Australis), perhaps under the influence of Aratus' comparison of it to a wreath here.

415–22 Aratus uses the Scorpion's tail (κ υ λ Scorpii) to guide the reader due south to the Altar, or Shrine (Ara), a small constellation at the limit of the southern sky as seen from northern latitudes. The core of the constellation is the quadrilateral α β ζ ϵ Arae. For an observer at 37° N the Altar would have been above the horizon for 4 1/2 hours, and Arcturus would have been below it for the same amount of time. Conversely,

Arcturus would have been above the horizon for 19 1/2 hours, and the Altar would have been below it for the same amount of time (Erren 1967, 66–67; Kidd 1997, n. 405). In myth the Altar is that upon which the Olympian gods swore an oath of allegiance before their war against the Titans.

423–56 Aratus explains that the Altar serves as a weather sign for seafarers. Though, as an offspring of primordial Chaos in Hesiod's *Theogony* (123), Night is indeed an "old, old" goddess, she here operates as an agent of the benevolent Zeus by pitying sailors and providing them with celestial signs. According to Aratus, it is normal for northerlies to pack thick clouds above the Altar in autumn. When the same phenomenon occurs in other seasons, one should expect stormy winds from the south. This series of warnings anticipates the alternatives that characterize later sections of the poem concerning weather signs.

At line 437 Aratus focuses on seafaring life and its risks. He does not specify the shoulder of the Centaur to which he is referring (450; see note on 456–63 below), and scholars dispute the nature of the "signals" which the Altar gives in 453. The most likely interpretation is as follows: when the brighter right shoulder of the Centaur (θ Centauri) reaches its zenith in the night sky but is obscured, and when the Altar itself is clear, one should expect winds from the east.

456–63 In the northern hemisphere the Centaur (Centaurus) lies along the southern horizon. Aratus states that his human head, shoulders (left ι Centauri, right θ) and arms (right ψ η κ) are under the Scorpion, and his horse's body (μ ζ ϵ τ), front hooves (α β), and hind legs (ρ δ π ο λ) are under the Claws. Aratus' directions are misleading, however: the constellation is under the Hydra's tail and the Claws. A line from the Centaur's right shoulder (θ Centauri) to right hand (η) points directly to the Altar. In myth this centaur (not to be confused with the centaur Sagittarius) is associated with Chiron, who trained the heroes Jason and Achilles.

The Beast is a constellation distinct from the Monster, or Sea-Beast (Cetus), that stalks Andromeda (note on lines 364–70, above). Ancient sources disagree as to what sort of creature this constellation represents. In myth it is a beast speared by the Centaur's thyrsus (ivy-bound staff), to which constellation it was attached. The Wolf (Lupus) is its modern name. η χ ψ^1 Lupi mark its head, and ζ its haunches.

463–71 Aratus uses the constellation Hydra, or Water-Snake, to introduce the Mixing-Bowl and Crow and the star Procyon. He does not specify stars in this constellation but points out its position relative to other constellations: its head (ζ ϵ δ σ η Hydrae) is under (south of) the Crab, its midsection (φ μ λ) is under the Lion, and its tail (π γ) is above (north of) the Centaur. The Mixing-Bowl (Crater) and the Crow (Corvus) lie on the Hydra itself, the former beneath the Lion and the latter beneath the Maiden. η θ Crateris delineate the Mixing-Bowl's lip and β α its base. A crater was a type of vessel used most often for mixing wine with water. The Crow consists of a quadrilateral of stars (δ γ ϵ β Corvi) for its body and a single star (α) for its head.

Procyon, a first-magnitude star, is important because its morning rising occurs at the hottest time of the year, anticipating the morning rising of Sirius. Hence its name, *pro* ("before") + *kuon* ("the dog"). For more on these risings, see Appendix 1. To the naked eye Procyon appears to be the brightest star in the constellation Canis Minor (not mentioned by Aratus). It is, however, a binary star system, consisting of Procyon A and Procyon B.

471–74 Aratus here concludes his descriptions of the northern and southern skies, emphasizing their regularity in order to set up a contrast with the erratic planets, which begin the next section.

The Planets and the Great Circles: 475–782

475–83 The ancient Greeks regarded the planets as stars, naming those that they saw the stars of Hermes, Aphrodite, Ares, Zeus, and Kronos—now known by Roman names as Mercury, Venus, Mars, Jupiter, and Saturn, respectively. These "stars" are of no use to Aratus because their movement is erratic. The sun and moon, however, do figure prominently in the sections of the poem concerning weather signs. The periods of the planetary years (the time it takes each to return to the same point in the Zodiac) differ widely and do not correspond to the sun's. Magnus Annus is the time when the sun, the moon, and all the planets conjoin (Plato includes the fixed sphere in this conjunction). Because Magnus Annus was held to mark the end of one era and the beginning of another, mathematicians and philosophers (especially Platonists and Stoics) attempted to determine the length of time from conjunction to conjunction. Aratus here employs *praeteritio*, a rhetorical device by which an author calls attention to a topic by the very act of dismissing it.

484–89 Aratus here systematically progresses towards his ultimate aim for the first section of the poem: judging the advance of night by the stars. Now that the addressee can recognize the constellations, he must also learn the four circles which surround the heavens. These four circles are theoretical. Two of them mark the furthest latitudes north and south in the sun's annual travels (23.5° N and 23.5° S, respectively), and one the midpoint between these latitudes (0°): these lines are the northern tropic (the celestial Tropic of Cancer), the southern tropic (the celestial Tropic of Capricorn), and the celestial equator, respectively. The fourth circle marks the track through which the sun travels in a year; it is called "the ecliptic" because solar eclipses can only occur when then moon crosses this track. The ecliptic is the most important of these lines for Aratus' purposes because the twelve Zodiac constellations lie along it.

490–500 The Milky Way is not one of the four circles Aratus mentions in the introductory section above (484–89); he introduces it here to indicate to the reader the length of the two broader circles—the celestial equator and ecliptic. As in other moments of excitement Aratus here uses the second-person singular pronoun: it is as if he were sitting and lecturing to a single pupil under the stars, rather than addressing a group. He does not, however, refer to himself in the first person, as he does in similar passages elsewhere.

500–522 The wind god Boreas here stands for a wind that issues from the north. The northern tropic (the celestial Tropic of Cancer) is the first of the two smaller circles. Starting with the Twins and proceeding westward to the Crab, Aratus lists prominent stars in constellations to trace this theoretical circle through the heavens: the Twins' heads (α β Gemini); the Charioteer's knees (between θ and β Aurigae); Perseus' left shoulder and left leg (θ ξ Persei); Andromeda's elbow (σ Andromedae); the Horse's front hooves (μ λ κ Pegasi); the Bird's neck (φ Cygni); the Serpent-Holder's right and left shoulders (β and κ Ophiuchi, respectively); the Lion's breast, belly, and genitals (α θ β Leonis, respectively); and the Crab's eyes (γ δ Cancri). Aratus here lists stars that are as much as 10° latitude to the north or south of the Tropic of Capricorn (23.5° N). Andomeda's elbow and the Horse's front hooves, however, do lie on this line.

By dividing this tropic into eighths, Aratus is using a rough geometric method for determining one's latitude on earth. For an observer at the equator, one-half (4/8) of the northern tropic is visible; for an observer at roughly 67° N (just north of the Arctic Circle) the entire northern tropic (8/8) is visible. As an observer north of the equator but far south of the Arctic Circle, Aratus sees 5/8 of the northern tropic.

523–34 Aratus traces the course of the southern tropic (the celestial Tropic of Capricorn) eastward from Capricorn itself, through the Water-Bearer's feet (δ Aquarii) and the Sea-Beast's tail (β Ceti), the small Hare (γ and ϵ Leporis, in particular), the Dog's hind paws (ϵ ζ Canis), the Argo (ρ ξ Puppis), the Centaur's back (just beneath the shoulders, which are θ ι Centauri), the Scorpion's sting (λ υ Scorpii), and the Archer's bow (λ Sagittarii). These stars are, in fact, on or near the Tropic of Capricorn (23.5° S).

On Aratus' division of this tropic into eighths, see note on 500–522 above. From his latitude above the equator Aratus sees 3/8 of the southern tropic and 5/8 of the northern.

534–47 Aratus proceeds to one of the two broader circles, the celestial equator. Twice a year the sun passes through this line; during the vernal and autumnal equinoxes the day and night are each twelve hours long. For guidance, Aratus cites the Ram (only a head for Aratus), the Bull's knees (ν Tauri), Orion's Belt (δ ϵ ζ Orionis), the Hydra's coils (ι Hydrae), the Mixing-Bowl (θ Crateris), the Crow (η Corvi), the Claws (β Librae), the Serpent-Holder's knees (ζ Opiuchi), the small Eagle (θ Aquilae), and the Horse's head (ϵ θ Pegasi) and neck (ζ ξ). Aratus again lists stars and constellations some distance from 0° latitude (the celestial equator). Orion's belt and the Hydra's coils, however, do lie very near the line.

548–58 Before moving on to the ecliptic, Aratus explains that the central axis piercing the earth (see lines 21–23) also supports the three parallel circles (the northern and southern tropics and the celestial equator). These theoretical lines are here discussed as if they were a physical object, such as the components of an armillary sphere. In this conception the "skewed" ecliptic serves to hold the three parallel bands on the outside, as the axis supports them from within.

"Athena's workinghands" is a Hellenistic metaphor for perfect craftsmanship (see Kidd 1997, n. 529). In the *Arogonautika* of Aratus' contemporary Apollonius of Rhodes, Aphrodite offers her son Eros a golden ball of a similar design:

> Hoops of gold bind the whole together; round them
> Parallel hoops, in turn, are sewn slantwise
> To cinch them tight; and blue streaks round the hoops
> In spirals wind and wander, covering all
> The seams and stitches. (3.135–40)

558–91 The twelve constellations which comprise the Zodiac all lie along the ecliptic; as the sun's annual course determines the ecliptic, the sun passes through them all each year. Half of the Zodiac is in view, half hidden under the horizon at any given time. In a mathematical image Aratus proposes a hexagon inscribed within the circle of the ecliptic; each side of the hexagon (60°) would contain two Zodiac constellations (roughly 30° each). A scholiast on this passage observes that the wax cells of bees' honeycombs are hexagonal, and the hexagon was regarded as one of the building blocks of the cosmos. Aratus here imagines the observer dividing up the Zodiac with his "eye-beams." The ancients assumed that the light which is reflected by eyes was emitted from them. For example, when Plato (428/427–348/347 BCE) argues that the soul is the source of vision, he cites light rays emitted from the eyes as evidence (Lindberg 1976, 1–17).

The following delightful doggerel mnemonic lists the Zodiac constellations in their modern order (starting with Ares the Ram):

The Ram, the Bull, the Heavenly Twins,
And next the Crab, the Lion shines,
The Virgin, and the Scales.
The Scorpion, Archer, and the Goat,
The Man who holds the Watering-Pot,
And Fish with glittering scales.

Aratus begins the list of the Zodiac constellations with the Crab, which the sun's path, or the ecliptic, reaches at the summer solstice (about June 22). Five constellations complete the first group of six: the Lion, the Maiden, the Claws, the Scorpion, and the Archer. Then comes Capricorn, which the ecliptic reaches at the winter solstice. Five further constellations complete the second group: the Water-Bearer, the Fish, the Ram, the Bull, and the Twins. Dividing the Zodiac into two groups of six underscores the point that half of these constellations (6/12) appear in the heavens at any time during the night.

Since dawn occurs during the risings of these constellations at fixed times each year, an observer is able to predict the length of time until dawn, the advance of the seasons, and, with less precision, related changes in the weather. When clouds cover the eastern horizon, however, the observer must rely on the position of the other Zodiac constellations to determine which one is rising (and thus how far a given night or the year has advanced). The "horns" of the ocean are the eastern and western hemispheres of the horizon.

591–617 At the morning rising of the Crab (Cancer) in the east, the Crown and the Plowman set entirely in the west. The setting of the Plowman takes a long time, spanning the risings of the Ram, the Bull, the Twins, and the Crab. His brightest star has its evening setting on October 30, marking the onset of winter, and the nights before this setting are therefore named after the Plowman.

Four constellations set partially: the Southern Fish up to its spine (ε λ Piscis Austrini), On-His-Knees up to the stars marking his lower abdomen (ε ζ Herculis), the

Serpent-Holder's head (α Ophiuchi) and shoulders (β γ ι κ), and the Serpent past its head (γ κ β Serpentis Capitis). Orion and the River rise with the Crab in the east—Orion as a sprawling form along the horizon up to his shoulders (α γ Orionis) and belt (δ ϵ ζ), and the River Eridanus entirely.

618–35 When the Lion rises in the east, the Southern Fish, the Serpent-Holder, the Serpent, and the Eagle set entirely, and On-His-Knees sets up to his left foot (ι Herculis) and hand (o). Along with the Lion in the east rise the Hare, Procyon, and the Dog's forepaws (β Canis Majoris).The rising of the Hydra spans the rising of four Zodiac constellations, the Crab, the Lion, the Maiden, and the Claws (Libra). Only the head of the Hydra (δ ϵ ζ η σ Hydrae) emerges with the Crab.

With the rising of the Maiden, the Lyre, the Dolphin, and the Arrow set entirely; the Swan's tail (α Cygni) and west wing tip (ζ Cygni) also set, along with the River up to θ Eridiani, and the Horse's head (θ ϵ Pegasi) and neck (α ξ ζ) with their trappings. At this time the Hydra continues to rise, lifting its coils above the horizon as far as the Mixing-Bowl, which rests upon them (past ν Hydrae to α β γ δ Crateris). The Dog's hind paws (ζ Canis Majoris) join its forepaws (β) above the horizon. The rear half of the Argo emerges as far as its central mast (γ λ Velorum).

635–67 With the Claws (Libra), the Plowman (containing the star Arcturus) and the Argo rise completely. The Hydra rises up to its tail (a point between β and γ Hydrae). On-His-Knees rises backward and upside down during the risings of the next three Zodiac constellations—the Claws, the Scorpion, and the Archer. With the Claws, On-His-Knees' right leg and knee (τ φ υ Herculis) appear in the east. Half of the Crown ascends (β up to α Coronae Borealis), along with the tip of the Centaur's tail (π Centauri). The Horse and the Swan now set completely, along with Andromeda's head (α Andromedae). As the Monster sets as far as its rear fin (β Ceti), Cepheus sets up to his head (δ ϵ ζ Cephei), right shoulder (α), and hand (θ).

668–712 When the Scorpion rises, the River, which rose with the Claws, sets. Orion also sets, and Aratus interprets this setting in mythological terms as a frightened retreat from the advancing Scorpion. In the earliest version of the myth Artemis kills Orion on Delos either because he challenged her in a discus competition or because he attempted to rape her or one of her maids, Opis. In Aratus' version the conflict is moved to the island of Chios because of Orion's friendship with Oinopion, the son of Dionysos and Ariadne, who founded a city on the island. Here Orion, in addition to clearing the island of beasts, also attempts to steal a piece of Artemis' clothing for him. (In another version Orion attempts to rape Oinopion's daughter.) As punishment for his offense against her, Artemis sends a scorpion to kill Orion.

When the Scorpion rises, Andromeda and the Monster, who had begun to set with the Claws' rising, now set completely. Cepheus (whose head is to the south) now sets to its furthest extent, his head (δ ϵ ζ Cephei) and sword-belt (β) now setting under the horizon. His feet and legs (γ κ) and loins (around β), however, are circumpolar and therefore never set. Here the tragic tone that has suffused Aratus' descriptions of the Cepheids turns paratragic, as Cassiopeia, flipped upside-down, dives head-first beneath the horizon. Aratus humorously suggests that this disgrace is a further punishment for her boast. Only the upward-tilted lower half of her constellation remains in view: her knee (δ Cassiopeiae) and foot (ϵ).

Nearly all the Crown rises at this time, as do the Hydra's tail (γ π Hydrae) and the Centaur's head (θ Centauri and above), along with the rest of his constellation up to his front hooves (α β). More of On-His-Knees emerges, including his right leg (η σ τ φ υ Herculis), shoulder (β), and hand (γ ω).

Aratus here anticipates the rising of the Bow (part of the Archer's constellation, handled in the next section, 712–30), mentioning that with it rise the Centaur's front hooves (α β Centauri). After Scorpion brings up the Serpent-Holder's head (α Ophiuchi) and the Serpent's head (β γ κ ι Serpentis Capitis), the Bow brings up the former's arms and more of the latter's coils (δ ϵ ν Ophiuchi, shared with the Serpent).

712–34 With the rising of the Archer (Sagittarius), the entire Lyre rises, and Cepheus' head (δ ϵ ζ Cephei) and breast (below β) emerge again (they had set with the rising of the Scorpion). Orion, the Dog, and the Hare now set entirely. The Charioteer sets during the risings of the Archer and Capricorn. Aratus here describes its setting in one passage rather than splitting the account up into separate ones: the lower part of his body sets first with the rising of the Archer; the upper part, consisting of head (δ Aurigae), right hand (θ), and waist (between θ and ζ), sets later with the rising of Capricorn. The Goat (α Aurigae), which also marks the Charioteer's shoulder, sets at this time with the Kids (ϵ ζ η Aurigae), which compose his arm. All of Perseus sets except his knee (μ Persei) and right foot (58R). As the Argo rises stern-first (Puppis) with rising of the Maiden, so its stern is the first part of the constellation to set here with the Archer rising. The rest (Carina, Vela, and Pyxis) sets when Capricorn rises. With Capricorn rising, Procyon, the Bird, the Eagle, the Arrow, and the Altar rise completely.

734–43 When the Water-Bearer (Aquarius) rises, the Horse's head (ϵ θ Pegasi) and front hooves (λ κ ι) rise. The Centaur's tail (π Centauris) and the Hydra's head (δ ϵ ζ η σ Hydrae) and neck (θ ι α) set. The rest of the Centaur and Hydra set when the Fish rise.

743–53 When the Fish (Pisces) rise, most of the Southern Fish rises, though a small part (υ ι Piscis Austini) awaits the rising of the Ram. After Andromeda's right shoulder (π Andromedae), forearm (σ θ κ), and knee (ν) rise with the Fish, her left side (γ δ) rises with the Ram.

754–59 When the Ram ascends, the Altar begins to set and Perseus' head (τ Persei) rises, with his chest—that is, the space between his shoulders (γ θ) and belt (α). Aratus explains that authorities differ over whether Perseus' belt (α) rises with the Ram or Bull, but the rest of his constellation certainly rises with the Bull.

759–70 The star β Tauri serves as both the Charioteer's right foot and the Bull's left horn-tip. As the two constellations are connected, the Bull brings up the Charioteer's right leg (θ) and left (ι Aurigae) at its rising. The Goat and the Kids, comprising the Charioteer's left shoulder (α Aurigae) and arm (ϵ ζ η), rise with the Bull. The Monster's tail (β Ceti) and rear fin (η β) also rise. The Plowman's setting spans the risings of four Zodiac constellations, the Bull, the Twins, the Crab, and the Lion (see 604–11 and note on 591–617 above). As it is circumpolar, the Plowman's left hand (θ ι κ Boötis) never sets.

771–80 When the Twins (Gemini) rise, the Serpent-Holder sets under the horizon up to his knees (η ζ Ophiuchi), and the Beast rises completely. The River Eridanus begins to rise (λ β ω μ ν Eridani). During the rising of the Twins, the observer can expect the large, prominent constellation Orion to rise soon with the Crab, the Zodiac constellation with which Aratus began his account of risings and settings (see 806–10 and note on 805–14 below).

781–82 These lines conclude the section of the *Phaenomena* dealing with the constellations and their use in assessing the advance of night. Some early manuscripts and translations draw a division between 1–782 and 783–1189, entitling the former the *Phaenomena* and the latter the *Diosemia* (Weather Signs) as if it were a separate work. The earliest manuscripts, however, make no such division; and although I include the heading here for convenience, I regard these two sections as parts of a whole.

Weather Signs: 783–1188

783–90 Aratus narrows his focus from the larger month-long changes of the twelve Zodiac signs to individual days marked out by the phases of the moon. Though nonlunar calendars had been introduced hundreds of years before Aratus' time in Athens and elsewhere, Aratus refers only to lunar months, as they arise from celestial phenomena.

To engage the addressee after the long and technical account of risings and settings, Aratus uses the second-person singular in a protreptic rhetorical question: "Don't you see?"—"Do you make out the following?" In Homer this phrase occurs at moments of emotional excitement and, most pertinently, when Penelope recognizes the omen in Telemachus' sneeze (17.545). The new moon with which the lunar month begins is itself difficult to see, since it appears only as a silhouette.

Aratus reckons the new moon as the first day rather than as zero, as we do, so his method of reckoning requires further translation: Aratus means the fourth day if one counts the new moon as the first, but this is only the third night of light on the moon, and thus the first eighth (3/28). Similarly, he means the first quarter when he refers to "eight days upon [the moon]"—this is only seventh night of light on the moon (7/28). The moon is full at midmonth (14/28).

791–804 Aratus has spent nearly two hundred lines (591–780) explaining how the annual progression of Zodiac signs can be used to determine the length of time before the end of a night; he here goes on to explain that stars and constellations can be used to determine specific days of the year. He cites the morning rising of Arcturus (Sept. 14) as an example and then draws a general conclusion: identification of the constellations rising at dawn and dusk allows one to prepare for upcoming seasonal changes in the weather (Geminos' *Calendar* in Aujac 1975). For a full account of the risings and setting of stars, and of the dates for Arcturus in particular, see Appendix 1.

805–14 Aratus mentions in passing the Metonic cycle, which reconciles lunar months with solar years by adding seven intercalary months over a nineteen-year period. Circa 430 BCE Meton and Euctemon incorporated the nineteen-year cycle into the *parapēgmata* in Athens. These public inscriptions on stones listed risings, settings, and key dates and served as a type of farmer's calendar. For an English translation of Euctemon's *parapēgma*, see Van Der Waerden 1984, 105–6. As the *Phaenomena* concerns only what is visible to the eye, Aratus does not refer to this cycle again, confining his account to lunar months and solar years.

The implied panorama of the night sky begins with Orion's morning rising and ends with Orion again. Orion rises with the Crab (612–16), which is the first Zodiac sign Aratus considers in his detailed account of risings and settings above (591–617). This account concludes with the rising of the Twins (771–80), when Orion's rising is anticipated (778). For more on this subject, see Appendix 1. In lines 809–10 Aratus refers to the sea and sky, by metonymy, as Poseidon and Zeus respectively. To avoid confusion I have opted to translate them in their figurative sense.

815–21 Aratus returns to the overarching purpose of the second section of the poem: to make the reader a more observant person. Before a storm, a ship reefs sails and stows or secures loose gear.

821–31 The benevolent Zeus is here introduced again. He has not only fixed the celestial bodies in the heavens to give us signs but also arranged a variety of further signs: changes in the appearance of celestial bodies such as the sun, the moon, and the Manger now become important, as does the behavior of flora and fauna. Furthermore, because Zeus is coextensive with the cosmos, he is immanent in all of these signs, as in all things. Aratus first concedes here that humans are not able to fathom Zeus' entire plan (821–23). Since the second section of the poem concerns signs that are not as regular and reliable as the movements of the stars, he further concedes (1177–79) that the conclusions one draws from weather signs are fallible and urges the observer to seek confirmation from as many as possible.

832–1175 Aratus classifies weather signs in two different ways: (1) by the location of the sign—the Moon (832–77), the Sun (877–936), and the Manger (937–53); (2) by the type of sign—Wind Signs (953–78), Rain Signs (979–1028), Signs of Fair Weather (1028–50), Signs of Foul Weather (1051–81), Signs of the Seasons (1081–99), and Animal Signs of Foul Weather (1099–1175).

832–860 The first part of this discussion of lunar signs focuses on the appearance of crescents—in particular, on the color and shape of the two tusks, or horns, at either end of them. At line 837 begins an acrostic hinging on the word *leptē* ("slender" or "elegant"), serving simultaneously to describe the slender tusks of the new moon's crescent (first visible on the third day) and as an expression of the Callimachean esthetic credo. For more on this acrostic, see the discussion of Hellenistic poetry in the Introduction.

A red moon is a visually striking phenomenon that occurs when a concentration of particles in the air (e.g., dust or smoke) scatters the short and intermediate wavelengths of light (violet, blue, and yellow) and only the longer wavelengths (orange and red) reach our eyes. In an adaptation of Homer's famous epithet, "rosy-fingered Dawn," Sappho refers to a "rosy-fingered Moon" (f. 96V). The inclination of the moon's horns in actuality has no connection to the weather: it depends upon the angle between the ecliptic and horizon. The "fourth night" here refers to the fourth night since the month began but the third phase in the cycle.

861–66 Aratus regards the lunar month as a meteorological unit in which certain key phases—the first eighth, the first quarter, the full moon, the third quarter, and the seventh eighth—forecast weather for a period of days but not the rest of the month (Kidd 1997, n. 805).

867–77 Halos first appear here and return later in discussions of the sun (923–26) and stars (985). According to Aristotle they are a reflection of the light emitted from our eyes on a smooth surface and occur when "air and vapor are condensed into a cloud, if the condensation is uniform and its constituent particles small" (*Meteora* 373a35, 372b16; see 372b18–34). Modern science has shown that halos result from the refraction of light when it passes through columnar ice crystals in high, thin clouds (i.e., cirrus or cirrostratus).

877–936 As most solar signs appear during sunrise and sunset, Aratus focuses on those two times. For the relationship between stars and sunrise and sunset, see Appendix 1. This passage is less formally organized than most others; Aratus seems to be consciously imitating the loose associative logic common in collections of folklore, such as similar passages in Hesiod's *Works and Days*.

887–923 In his injunction and warning about looking into the sun, Aratus suggests the danger both of staring at the sun and of confronting a sun god. In myth, when presented with a choice by the sun god, Helios, between sight and a short life or blindness and a normal length of life, the prophet Phineus chooses blindness along with a normal length of life.

In lines 894–97 one would assume that the sun's beams appear concentrated because they are shining through a gap in surrounding cloud cover. The "arching clouds" besetting the sun are complete cloud cover.

At line 909 the sun is described as setting in two stages—descending toward the horizon and disappearing beneath it. In 917–23 Aratus explains that a murky sunrise can be a sign of either wind or rain: if the sun is darkly obscured, expect rain; if only lightly, expect wind.

923–36 For halos in general, see the note on 867–77 above. A "parhelion" (from the Greek, meaning "beside the sun"; plural *parhelia*) results from the same light refraction as halos. Parhelia usually appear in pairs at far ends of the halo arcs alongside the sun when it is rising or setting. They are also known as sundogs. There is no connection between the appearance of parhelia and wind.

937–53 The Manger is a star cluster (M44) in the middle of the Crab constellation (Cancer). It is flanked by two stars called the Asses (γ δ Cancri). According to Aratus, fluctuations in the comparative brightness of the Manger, along with that of the each Ass, allow an observer to predict storms, winds, and even the direction from which the wind will come. At 944–45 he refers to an optical illusion; the Asses only appear closer together when the Manger is not visible between them.

953–78 This beautiful passage lists miscellaneous wind signs taken from the sea, maritime birds, clouds, thistledown, lightning, and meteors. Aratus artfully arranges his birds so that the heron that flies with the wind contrasts with the petrel that faces it. The *kepphoi* (which I have translated as "storm petrels") are sea birds noted for their stupidity (Aristophanes *Peace* 1067). They have proved difficult to identify with certainty (see Thompson 1936, 137–38, and Kidd 1997, n. 916). In an act of anticipatory sympathy with the incoming wind, the wild ducks and shearwaters cause a disturbance by beating their wings on the beach. For the shearwater, see the note on 293–308 above. The name for the hoary down of an older thistle plant is *pappos* in Greek, which also is the word for "grandfather." This down contains the seed which, according Aratus, is released prior to the onset of high wind, presumably for dissemination abroad.

Aratus points out that wind (and probably a thunderstorm) will come from the direction in which one sees lightning. Kidd (1997, n. 926) traces the origin of Aratus' meteor imagery back to a simile describing Athena's descent from Olympus in Homer's *Iliad*, 4.75–77:

> The child of Crooked-Minded Kronos [Athena, daughter of Zeus]
> Flew like a star or bright portent for sailors or a broad army of men.
> And many sparks shoot forth from it.

Since meteors, or "shooting stars," result from matter that heats up due to fiction upon entering the atmosphere, they have no connection to the weather. In popular belief, however, the connection was taken for granted, and Aristotle cites the widely recognized coincidence of shooting stars and drought as evidence for the origin of meteors and comets in the hot, dry highest stratum of the air (Aristotle *Meteora* 1.6–7). See note on comets and meteors, 1098–1174 below.

979–1028 As Aratus collected miscellaneous signs of wind in the preceding passage, he now collects various signs of rain in four categories: the sky (979–86), littoral creatures (986–98), terrestrial creatures (999–1016), and domestic life (1017–28).

979–86 Zeus more frequently appears as a hostile weather-god in the sections concerning weather signs, and Homer describes Zeus' rainbow as a dire omen: "as when Zeus extends a purple rainbow to mortals to be an omen of war or a rough storm" (*Iliad* 17.547–49). A rainbow occurs when the sun shining through droplets of water in the atmosphere causes a spectrum of light to appear. On those rare occasions when a second, fainter rainbow appears, the colors are arranged in the opposite order, with violet along the top and red along the underside, as Aristotle had already observed (*Meteora* 375a30).

989 Aratus travesties Homeric language, inverting the patronymic formula by naming frogs in terms of their offspring ("fathers of tadpoles") instead of their progenitors.

1001–2 The ants are, in fact, hauling pupae, not eggs, out of their hole.

1003 "Earth's entrails" was a common name for worms (see Aristotle's *On the Gait of Animals* 705a27, *On the Generation of Animals* 762b26, and *The History of Animals* 570a16; and Pseudo-Theophrastus' *Diosemeia* 42).

1005 In another Homeric patronymic, fowl are referred to as "scions of the cock."

1010–13 The ravens here give us the clearest example of anticipatory sympathy in nature: they imitate the sound of the coming rain. It is striking that Aratus repeatedly uses the Greek word *phōnē* ("voice," in particular that of humans) in his descriptions of the fowl (1004–8) and ravens (1010–13). Their utterances are similar to human speech in that they convey meaning—the approach of rain or wind (Kidd 1997, n. 967).

1018 "Snuff" here is the charred portion of a candlewick.

1025 A tripod is a three-legged implement that can be set over a fire for cooking purposes.

1026 Along with grains of sand, millet seeds, being tiny and plentiful, were commonly used in comparisons to express anything infinitely numerous.

1028–50 Aratus devotes less space to the signs of fair weather than to the signs of foul because fair weather does not require precautionary measures. On the Manger see the note on 937–53 above. The "steady light" of a candle here is the opposite of the erratic flame given as a sign of rain (1019–21). On the basis of its behavior here, the owl has been identified as the little grey owl (*Athene nocturna*) (Thompson 1936, 76–78). This owl was depicted on ancient Athenian coinage and currently appears on the obverse of the Greek one-Euro coin. Its modern Greek name, Koukouvaya, is onomatopoeic.

1051–80 Aratus lists signs of foul weather found in stars (1051–55), clouds (1055–57), birds (1058–66), bees (1066–69), cranes (1069–70), cobwebs (1070–72), and candles and the domestic hearth (1072–80).

The cloud that is "swirled up in the air" (1056) is the tall, dense cumulonimbus cloud (or thunderhead) involved in thunderstorms and, at times, tornados. In folklore the crow is assumed to live for nine human generations (Hesiodic *Precepts of Chiron* fr. 3.1–2). The Homericizing adjective *enneagera* ("nine-times-old") raises the register, perhaps to encourage comparison of the crow with long-lived mythic wise men (e.g., Nestor who lived for three human generations, and the prophet Tiresias, who lived for seven). In Ovid's *Metamophoses* 7.273–74, Medea mixes "the eggs and head of a nine-lived crow" into the witch's potion that rejuvenates her husband Jason's father Aeson. The European robin (*Erithacus rubecula*) nests anywhere it can find a depression or hole. The wren here must be *Troglodytes troglodytes*, called the winter wren in North America and simply the wren in the Old World. It is the only Old World species of wren, and its genus and species names come from the Greek roots *trōglē* ("hole") and *dyein* ("to go down").

Aratus concludes the section on the signs of foul weather with the rhetorical question "But why . . . " (1074) to drive home the point that the signs of foul weather are numerous and he is only giving a few examples.

1080–98 This passage focuses exclusively on flora, in contrast to the animal, domestic, meteorological, and celestial phenomena presented in the other sections of "Weather Signs." The holm oak (*Quercus ilex*) is a large evergreen oak native to the Mediterranean region. It is also known as holly oak and evergreen oak. Mastich (*Pistacia lentiscus*) is an evergreen shrub or small tree of the pistachio family, noted for its aromatic resin. Though it is native throughout the Mediterranean region, the Greek island of Chios is famous for its cultivation. Squill is the common name for two genera of Old World bulbous plants in the Liliaceae (lily) family. For horticulturists, squill is any plant in the genus *Scilla*, which mostly bloom in spring. The common squill is the wild hyacinth (*Scilla nonscripta*), also known as wood hyacinth, bluebell, or harebell. For pharmacists, squill is the sea onion (*Urginea maritime*), which produces cream-colored or red flowers in autumn before producing leaves. None of these species has been observed to flower three times in a year, as Aratus claims; his observation may result from the observation of different flowering times among the various species of squill.

1098–1174 In this passage the behavior of farm animals (including wasps and even hermit crabs) predicts the early vs. late onset and departure of winter (and the storms that come with it); comets serve to predict a dry vs. a wet summer growing season.

1098–1129 The word that I have translated here as "Fall" (1098) is *metopwrinon*, or *meta* ("after") + *opwra* ("ripening"). The morning rising of Arcturus in mid-September was traditionally held to begin this season. With "Even before the Pleiades descend" in 1102, Aratus refers to the morning setting of the Pleiades in early November, as he does again with "the setting Seven Sisters" (1119). See the note on the Pleiades, 260–72 above, and Appendix 1 for a full account of constellation risings and settings.

In an act of anticipatory sympathy the wasps' hectic swarming imitates the fervor of the storm. The cranes are migrating south for the winter, an image which appears in both Homer (*Iliad* 3.3–4) and Hesiod (*Works and Days* 448–51). At 1125 the "stalks all are indistinguishable" because the grain is just sprouting, and one cannot yet discern separate stalks.

As comets come only at intervals of many years, it would be a very rare occurrence for more than one to appear in a single year. Aratus, however, regards meteors (a much more frequent occurrence) and comets as the same phenomenon. Aristotle, in fact, classifies meteors, or shooting stars, as a kind of comet (Aristotle *Meteora* 1.6–7).

1130–40 The migrating birds that are a bad sign for the farmer are a good sign for the goatherd. In a striking act of sympathy so close to the end of the poem, Aratus' narrative voice gives up its authoritative distance and lumps the poet in with the rest of humanity by using a first-person plural pronoun (we): though different men have different careers, we are all alike in our pursuit of prosperity and our reliance on predictive signs.

1141–62 Aratus sets this series of signs firmly within the world of the pastoral, which his contemporary Theocritus is credited with establishing as a distinct genre. A lone wolf was considered to be particularly desperate and destructive. Demosthenes is said to have referred to the young Alexander the Great as the "the lone wolf from Macedon" (Plutarch *Demosthenes* 23.4).

1163–74 In this final litany of weather signs Aratus again elevates humble animals (mice, dogs and hermit crabs) with epic themes (the customs of forebears) and personification (dancing). At this point, however, the elevation reads less as mock loftiness than the proper register for creatures that are filled with Zeus and whose behavior conveys his signs. The word here for "squeak," *tetrigotes*, is also used of the shades of the dead and of bats (Homer *Odyssey* 24.5 and 7).

1175–88 Providing parting advice regarding the star calendar (1178–80) and lunar phases (1180–86), the conclusion brings material from the two sections of the poem together. Aratus frames this advice between injunctions (1175–78, 1186–88) to look for a variety of weather signs and to place one's confidence in the greatest coincidence of them.

Works Cited

Auden, W. H., and Pearson, N. H. 1953. *Poets of the English Language.* Vol. 3. The Viking Portable Edition. New York: Viking Press.

Aujac, Germaine, ed. and trans. 1975. *Géminos: Introduction aux phénomènes.* Paris: Les Belles Lettres.

Bing, Peter. 1990. "A Pun on Aratus' Name in Verse 2 of the *Phainomena?*" *Harvard Studies in Classical Philology* 93:281–85.

——. 1993. "Aratus and His Audiences." *Materiali e discussioni per l'analisi dei testi classici* 31:99–109.

Courtney, Edward. 1993. *The Fragmentary Latin Poets.* Oxford: Clarendon Press.

Effe, Bernard. 1977. *Dichtung und Lehre: Untersuchungen zur Typologie des antiken Lehrgedichts.* Munich: Beck.

Erren, Manfred. 1967. *Die Phainomena des Aratas von Soloi: Untersuchungen zum Sach- und Sinnverständnis.* (Hermes: Zeitschrift für Klassische Philologie, Einzelschriften 19.) Wiesbaden: Franz Steiner.

Fakas, Christos. 2001. *Der hellenische Hesiod: Arats Phainomena und die Tradition der antiken Lehrepik.* Wiesbaden: Reichert.

Fantuzzi, Marco, and Hunter, Richard. 2004. *Tradition and Innovation in Hellenistic Poetry.* Cambridge: Cambridge University Press.

Fowler, Barbara H. 1989. *The Hellenistic Aesthetic.* Madison: University of Wisconsin Press.

Gain, D. B. 1976. *The Aratus Ascribed to Germanicus Caesar.* London: Athlon Press.

Gee, Emma. 2000. *Ovid, Aratus, and Augustus: Astronomy in the "Fasti" of Ovid.* Cambridge: Cambridge University Press.

Gutzwiller, Kathryn. 2007. *A Guide to Hellenistic Literature.* Oxford: Blackwell Publishing.

Herbert-Brown, Geraldine. 1994. *Ovid and the "Fasti"* Oxford: Clarendon Press.

Herzog, Reinhart. 1993. *Nouvelle histoire de la litterature latine.* Brepols: Turnhout.

Hinds, Stephen. 1987. *The Metamorphosis of Persephone.* Cambridge: Cambridge University Press.

Honigmann, Ernest. 1950. "The Arabic Translation of Aratus' Phaenomena." *Isis* 41:1, 30–31.

Hunter, Richard L. 1995. "Written in the Stars: Poetry and Philosophy in the *Phaenomena* of Aratus." *Arachnion* no. 2.1 (September), at www.cisi.unito.it/arachne/num2/hunter .html.

Hutchinson, G. O. 1988. *Hellenistic Poetry.* Oxford: Oxford University Press.

Kidd, Douglas, ed. and trans. 1997. *Phaenomena.* Cambridge: Cambridge University Press.

——. 1981. "Notes on Aratus, *Phaenomena.*" *Classical Quarterly*, n.s., 31:355–62.

Le Bourdellès, Hubert. 1985. *L'Aratus Latinus: Études sur la culture et al langue latines dans le Nord de la France au VIIIe siècle.* Lille: University of Lille.

Lefkowitz, Mary R. 1981. *The Lives of the Greek Poets.* Baltimore: Johns Hopkins University Press.

Levitan, William. 1979. "Plexed Artistry: Aratean Acrostics." *Glyph* 5:55–68.
Lindberg, David C. 1976. *Theories of Vision from Al-Kindi to Kepler*. Chicago: University of Chicago Press.
Long, A. A. 1974. *Hellenistic Philosophy*. London: Duckworth.
Sachau, Edward C. 1910. *Alberuni's India*. 2 vols. London: Kegan Paul, Trench, Trübner.
Selden, D. L. 1998. "Alibis." *Classical Antiquity* 17:289–412.
Shiler, E. G. 1933. *Cicero of Arpinum*. New York: G. E. Stechart.
Thompson, D. W. 1936. *A Glossary of Greek Birds*. 2nd ed. London: Oxford University Press.
Townend, G. B. 1965. "The Poems." In *Cicero*, ed. T. A. Dorey. London: Routledge and Kegan Paul.
Van Der Waerden, B. L. 1984. "Greek Astronomical Calendars, I: The 'Parapegma' of Euctemon." *Archive for History of Exact Sciences* 29, no. 2: 101–14.
Volk, Katharina. 2002. *The Poetics of Latin Didactic: Lucretius, Vergil, Ovid, Manlius*. London: Oxford University Press.